HISTORY

OF THE

RISE AND PROGRESS

OF THE

IRON TRADE

OF THE

UNITED STATES,

FROM 1621 TO 1857.

WITH NUMEROUS STATISTICAL TABLES, RELATING TO THE MANUFACTURE, IMPORTATION, EXPORTATION, AND PRICES OF IRON FOR MORE THAN A CENTURY.

BY B. F. FRENCH,

MEMBER OF THE GEOGRAPHICAL AND STATISTICAL SOCIETY OF NEW YORK, AND THE ACADEMY OF NATURAL SCIENCES OF PHILADELPHIA, ETC., ETC.

NEW YORK:
WILEY & HALSTED.
1858.

W. H. TINSON, PRINTER AND STEREOTYPER.

TO

PROFESSOR ALEXANDER D. BACHE,

SUPERINTENDENT OF THE U. S. COAST SURVEY,

THIS VOLUME IS DEDICATED,

TO EXPRESS MY ADMIRATION OF HIS

EMINENT SCIENTIFIC AND LITERARY ATTAINMENTS,

AND THE HIGH VALUE I ATTACH TO HIS

FRIENDSHIP.

BENJ. F. FRENCH.

No. 94 Clinton Place, New York.

INTRODUCTION.

In preparing this volume for publication, it has been my object to furnish a reliable work upon the Iron Trade of this country—showing its past and present condition, the obstacles it has had to contend with, and the effect of the free-trade principles which have crept into our legislation, and produced those finacial revulsions which have, from time to time, brought ruin upon this branch of American industry.

There is no article of manufacture more important than iron—none in which it is more necessary that a nation should be independent of all the chances of war or commercial restrictions—the large employment of which is the chief characteristic of civilization, and should always be, if possible, a domestic production, and protected beyond the possibility of any contingency.

It is, no doubt, difficult to lay down any general rule as to the rate of protective duties. They should, however, be more than sufficient, in all cases, to protect the manufacture of any article which it is the design of a foreign government, foreign merchants, or foreign manufacturers to crush. It is an established fact, that, in many departments of English industry, those who are interested will not only carry them on at a loss for years, but furnish large sums of money besides to aid in retaining markets from which they are in danger of being excluded by commercial restrictions or industrial competition ; and scarcely a branch of industry has sprung up in the United States which has not had, at first, to encounter a severe struggle, by the reduction of the price of the foreign article, below that which the new manufacture was expected to compete with.

It is, therefore, premature for this country to adopt a free trade policy, while her manufactures are yet undeveloped, and unable to compete with foreign countries. "It may be a good policy," says a recent writer in one of our daily journals, "for some countries; and is, where the manufacturing interests are fully developed, and where there is a dependence for the raw material and for food upon other countries, as in the case of a country like Great Britain, and at some future day, probably, our own; but the important question is, whether we are *now* in that situation—whether our manufacturing interests in their infant state can compete with the cheap capital and labor of Europe, and whether our iron, copper, and lead ores shall always remain unworked, or be made, by native industry, to give wealth to our country? Or, again: Can we, with all our resources, afford to support decrepit Spain, add wealth to Cuba, aid France and Great Britain in their thousand per cent. *ad valorem* tobacco duties, and build up Brazil with our "free trade?" To show that we are not indulging in wild assertions, let us prove these statements by statistics. The balances of trade with some of these countries, for the year 1857, appear thus:

COUNTRIES.	Exports from U. S.	Imports into U. S.	Balance of Trade against U. S.
Brazil, ...	$5,545,207	$21,400,733	$15,915,526
Spain (including Cuba), ...	27,889,568	57,432,545	29,542,977
Mexico, ...	3,615,206	5,985,857	2,370,651
Venezuela, etc., ...	3,502,443	6,344,490	2,842,047
Total, ...	$40,552,424	$91,163,625	$50,671,201

"From this statement it appears, that in 1857 the balance of trade against the United States in its trade with four neighboring countries, was over fifty millions of dollars for the year 1857; or the imports were 70 per cent. of the total commerce. These countries are chosen as an example, from their proximity to this country, and not solely from the great balance of trade against us. If China, etc., had been included, the balance would have been much greater. These countries having close proximity to the United States, the annual trade with them should be large, and, being in different latitudes, there is a mutual necessity and demand for the productions and manufactures foreign to the respective climates; and consequently there would be with 'free trade,' or *equal tariffs*, an approximately equal trade. Now, as we cannot get 'free trade' from these countries, we should have equal tariffs. We also find that the balance of trade with these four countries *absorbs* the entire gold productions of California, and that we contribute that amount annually to the wealth of these countries. Here is a fact for free-traders to consider!

"An equality of trade is, in a great measure, dependent upon an equality of tariff; and an equality of trade, as a general rule, is the only true basis of commerce. One illustration will prove this. The duty upon a barrel of flour entering Cuba will average about ten dollars, or two hundred per cent. of its value. Now, flour is a staple product of the United States, and a necessary of life in Cuba, and from the proximity of the two countries, a large trade in this article would be a legitimate result; but owing to the tariff duty, the trade is merely nominal; and although generally there is a greater demand in Cuba for the products of this country than the reverse, yet with Cuba the ratio of our imports to exports is *three to one,* giving an annual balance of trade against us of almost eighteen millions of dollars.

"Is it not, then, proved that the 'revenue' is not the only element to be considered in a tariff—protection is also necessary?

"One of our principal staples is tobacco; the crop of this year, 1858, being valued at twenty-five millions of dollars. We are the largest producers of tobacco in the world, and supply the non-producing countries of Europe almost entirely. Yet, what are the facts in regard to tobacco? France obtains a yearly revenue of nearly thirty millions of dollars from tobacco, or five millions more than the value of our entire crop. We supply France with four-fifths of her consumption, and it is estimated that, with a moderate duty, our exports would increase ten-fold. The revenue to Great Britain on tobacco is over twenty-two millions of dollars annually; of Holland, twenty millions of dollars; Austria, seven millions; Spain, five millions—and so through the list, showing how the monopoly of one of our staples supports some of the governments of Europe.

"That tobacco is not an exceptional case, let us examine the tariffs of Europe, in regard to duties levied on our leading staples.

Wheat.

Spain, ...	Prohibited.
Belgium, ...	19 cts. per 200 lbs.
France, ...	Duty fixed monthly.
Holland, ...	121 cts. per ton.

Cotton.

Spain,	1 to 2 cts. per lb.
Belgium,	77 cts. per 100 lbs.
France,	$2 44 to $3 42 per 100 lbs.

Cotton Manufactures.

Spain,	31 cts. to $4 67 per 100 lbs.
Belgium,	$34 to $57 per 100 lbs.
France,	Prohibited.

Provisions.

Spain,	6 cts. per lb.
Belgium,	96 cts. per 220 lbs.
France,	$3 55 per 100 lbs.

"These examples are, perhaps, enough to show how little 'free trade' exists in Europe, in regard to our staple productions; and the monopolies and exorbitant duties are even more common in South America, and in the few Asiatic countries with which we have dealings."

When the system of reciprocity and free trade had been proposed to a French ambassador, his remark was, "that the plan was excellent in theory, but, to make it fair in practice it would be necessary to defer the attempt to put it in execution for half a century, until France should be on the same footing with Great Britain, in marine, in manufactures, in capital, and the many other peculiar advantages which it now enjoyed." It was, truly, a profound remark, worthy the representative of a great manufacturing and commercial nation, and should not be lost upon this country.

In regard to the manufacture of iron in the United States, there is nothing required but *permanent*, secure protection, to afford every branch of iron manufactures cheaper than they could be obtained from England, or from any other foreign country. Many of them had already begun to be cheaper under the tariff of 1842. It requires time, confidence, and capital to perfect all the manufactures of iron ; and, while advancing to perfection, the prices should be satisfactory, beneficial to the country, and beneficial to all parties.

If, however, on the other hand, they should be discouraged, and suffered to languish for want of protection, England will most assuredly avail herself of the occasion to raise her prices higher than she would under an adequate system of protection.

CONTENTS.

HISTORY

OF THE

IRON TRADE OF THE UNITED STATES.

THE United States, since the acquisition of California and New Mexico, yield to no country in the richness and purity of her metallic ores, especially those of iron, which run in gigantic measures throughout the Union, and exist nowhere more abundantly or in greater variety than in the Carolinas, Tennessee, and Georgia.

These rich and inexhaustible deposits may be worked with great profit in every locality, and would soon become a source of great national wealth, but for the jarring interests of sectional feeling which prevent their development, and deny to them the preponderance to which their manufacture is entitled.

First, there is the great magnetic region which lies imbedded in the Appalachian chain of mountains, covered with vast and rich forests of timber, and at the base and along the valleys of which are found the brown and yellow hematite and pipe ores, scattered in rich abundance, reposing upon their limestone beds, and in juxta-position with all the materials necessary for their manufacture.

Secondly, the fossiliferous deposits, opened out extensively in Pennsylvania, Maryland, Tennessee, and Virginia, and also in the midst of immense forests of timber.

Thirdly, the argillaceous ores of the coal measures, so precious from their association with the bituminous coal deposits of this country ; and,

Fourthly, the iron mountains of Missouri and Michigan, which yield the purest of iron, and any one of which localities could furnish iron enough to supply the world.

The North American colonies engaged in the manufacture of iron at a very early period. In 1621, Virginia led the way, and was followed by Massachusetts in 1628. They made, however, but little progress, as the mother country adopted the policy of restricting their manufacturing spirit by administrative measures. In 1660, the British Parliament passed an act prohibiting the American colonies from exporting any of their manufactures to Great Britain in any other than English-built ships, although in direct violation of the Charter of Virginia, which empowered the people of that colony to carry on a direct trade with foreign countries.

In 1679, Great Britain imposed a duty of ten shillings per ton on all iron imported into the American colonies. It was afterwards proposed by the House of Lords, to prohibit the American colonies from manufacturing iron wares of any kind out of sows, pigs, or bars, under a heavy penalty, which did not, however, become a law. In 1731, an act was passed by Parliament directing the Board of Trade to inquire into and report on the laws made, manufactures set up, and trade carried on by the American colonies. In the following year, they accordingly reported that iron works had been for many years established in Massachusetts, Rhode Island, Connecticut, New York, Pennsylvania, Maryland, and Virginia, and from the progress they had made, it was expedient to encourage the manufacture of it in the colonies, especially as the production of it had greatly fallen off in the mother country.

The officers of several of his Majesty's dockyards, also reported that the American iron was *superior* in every respect to the best Swedes iron.*

* Copy of a letter from the officers of his Majesty's yard at Woolwich to the Navy Board, dated 3d September, 1735.

"We have lately received fron his Majesty's yard at Deptford, bar iron 2¼ inches broad and 1½ inches thick—15 cwts. 0 qrs. 4 lbs.; squares of ⅞ of an inch—5 cwts. 0 qrs. 12 lbs. imported by Mr. Crawley from America, and pursuant to your warrant of July 11th, 1735, have made sufficient trial of each of the sorts, find the said iron to be very good, and fit for his Majesty's service, *superior* in every respect to the best Swedes iron, and in our opinion worth £17 10s. 6d. per ton." They also wrote to the Navy Board, on the 17th Feb., 1736, "That, from the trials we have made from one ton of iron, imported by Mr. Crawley, from America, it is, both in the nature, and goodness, and value, equal, in all respects, with Swedes iron."

WROUGHT AND BAR IRON IMPORTED BY THE AMERICAN COLONIES FROM 1710 TO 1735.

	COLONIES.	IRON, WROUGHT.			BARS.			
		C.	Qr.	lb.	T.	C.	Qr.	lb.
1710 to 1711.	Carolina,........	1,143	0	27	00	00	0	00
	New England,....	4,596	2	16	200	3	0	7
	New York, ...	567	0	19	10	2	1	10
	Pennsylvania, ...	937	2	00	12	0	2	21
	Virg'a and Mary'd,	3,014	0	8	1	10	1	1
1712.	Carolina, ...	1,551	0	7	4	13	0	00
	New England,....	5,344	3	24	281	13	3	19
	New York, ...	639	1	7	32	3	0	00
	Pennsylvania, ...	540	0	20	2	1	0	00
	Virg'a and Mary'd,	5,653	2	4	5	3	2	14
1713.	Carolina, ...	1,406	2	2	27	5	0	00
	New England,....	983	0	13	212	1	0	00
	New York, ...	4,885	2	21	49	8	2	16
	Pennsylvania, ...	1,040	0	9	7	4	3	26
	Virg'a and Mary'd,	2,859	2	21	8	5	2	4
1714.	Carolina.	1,051	1	18	8	10	0	00
	New England,....	4,633	0	3	279	6	3	00
	New York, ...	1,136	3	15	98	7	0	18
	Pennsylvania, ...	923	2	1	24	12	0	7
	Virg'a and Mary'd,	6,597	2	12	8	5	0	00
1715.	Carolina, ...	691	0	21	1	18	0	00
	New England,....	5,795	2	24	372	10	1	00
	New York, ...	1,379	3	00	110	10	0	00
	Pennsylvania, ...	987	3	4	8	5	0	00
	Virg'a and Mary'd,	8,946	3	15	16	7	0	00
1716.	Carolina, ...	670	1	7	00	00	0	00
	New England,....	5,397	2	2	372	10	3	00
	New York, ...	1,094	0	14	147	00	0	00
	Pennsylvania, ...	962	2	00	10	00	0	00
	Virg'a and Mary'd,	7,446	0	22	8	10	0	00
1717.	Carolina, ...	866	1	11	4	2	0	00
	New England,.....	3,819	6	5	140	7	0	00
	New York, ...	1,146	1	00	42	14	1	00
	Pennsylvania, ...	1,149	0	26	8	16	2	00
	Virg'a and Mary'd,	8,728	1	27	3	3	0	00

	COLONIES.	IRON, WROUGHT.			BARS.			
		C.	Qr.	lb.	T.	C.	Qr.	lb.
1718.	Carolina, ...	970	2	21	2	00	0	00
	New England, ...	3,110	1	1	154	4	0	00
	New York, ...	1,396	1	20	2	18	2	00
	Pennsylvania, ...	887	0	2	3	10	0	00
	Virg'a and Mary'd,	6,734	2	3	26	10	3	00
1729.	Carolina, ...	1,342	1	21	3	10	0	00
	New England, ...	7,393	3	00	337	12	2	23
	New York, ...	1,903	2	23	58	01	0	25
	Pennsylvania, ...	851	0	14	4	00	0	00
	Virg'a and Mary'd,	4,866	0	23	1	1	0	00
1730.	Carolina, ...	1,479	3	23	5	10	0	00
	New England, ...	7,329	2	24	149	13	1	5
	New York, ...	2,775	0	00	91	10	2	13
	Pennsylvania, ...	2,628	3	00	00	00	0	00
	Virg'a and Mary'd,	6,389	2	24	2	9	0	00
1731.	Carolina, ...	1,770	0	11	10	18	0	7
	New England, ...	9,727	1	7	243	8	2	7
	New York, ...	2,627	2	7	101	11	1	00
	Pennsylvania, ...	2,946	0	9	5	00	0	00
	Virg'a and Mary'd,	9,681	3	11	3	18	0	00
1732.	Carolina, ...	2,167	3	7	9	00	0	00
	Georgia, ...	291	0	00	00	00	0	00
	New England, ...	8,597	2	4	413	5	2	00
	Pennsylvania, ...	2,207	2	26	2	14	0	00
	Virg'a and Mary'd,	7,445	3	27	4	14	0	00
		PIG.			BARS.			
		C.	Qr.	lb.	T.	C.	Qr.	lb.
1733.	Carolina, ...	2,692	3	11	25	00	0	00
	New England, ...	7,104	3	17	378	4	0	00
	New York, ...	1,609	3	7	55	00	0	00
	Pennsylvania, ...	2,419	2	0	2	00	0	00
	Virg'a and Mary'd,	8,815	1	10	12	00	0	00
1734.	Carolina, ...	2,880	2	19	25	00	0	00
	Georgia, ...	332	1	00	00	00	0	00
	New England, ...	7,104	3	14	370	14	2	00
	New York, ...	1,609	3	7	55	00	0	00
	Pennsylvania, ...	2,419	2	8	2	00	0	00
	Virg'a and Mary'd,	8,815	1	10	12	00	0	00

COLONIES.		PIG.	BARS.
		C. Qr. lb.	T. C. Qr. lb.
1735.	Carolina, ...	3,353 1 23	5 19 0 00
	Georgia, ...	1,700 0 00	4 00 0 00
	New England, ...	6,543 2 23	107 9 3 00
	New York, ...	2,136 2 00	108 8 1 5
	Virg'a and Mary'd,	9,709 1 24	2 13 0 00

A controversy sprung up in Great Britain about this time, on the policy of permitting American iron to be imported into that country free of duty.

It was maintained by the British merchants that, inasmuch as the importation of foreign iron into England was of great amount, and came mostly from Russia and Sweden, the principal part of which had to be paid for in money, and since it had been ascertained that the American iron, from the colonies, was equal in quality to either, and could be paid for in British manufactures, besides giving employment to British shipping, it was the interest of the mother country to encourage the manufacture of iron in the American colonies.

The English iron-masters, on the other hand, contended that to encourage the manufacture and importation of American iron free of duty into England, would tend to injure the English manufacturer, and reduce many thousand families to want.

The arguments of the English merchants, however, at length prevailed, and a law was passed by Parliament in 1750 permitting both pig and bar iron to be imported into London, from the American colonies, free of duty; but, at the same time, it prohibited them from erecting any rolling or slitting-mill, or forge to work with a tilt hammer, and from manufacturing steel for exportation, under a penalty of £200. The duty on bar iron thus taken off was £2 1*s.* 6$\frac{3}{20}$*d.* per ton, and on pig iron 3*s.* 9$\frac{9}{20}$*d.* per ton. In the following year, the Governors of the several American colonies reported to Parliament that there were in operation in the colonies four mills or engines for slitting or rolling iron, eleven plating forges to work with tilt hammers, and five furnaces for making steel.

PIG AND BAR IRON EXPORTED TO GREAT BRITAIN BY THE AMERICAN COLONIES FROM 1728 TO 1748.

COLONIES.		LONDON.	OUTPORTS.
		T. C. Qr.	T. C. Qr.
1728–9..	Pennsylvania, ...	99 13 2	174 12 3
	Virg'a and Mary'd,	344 9 0	508 6 3
1730....	Pennsylvania, ...	105 16 0	83 00 0
	Virg'a and Mary'd,	779 17 0	746 17 0
1730–1..	Pennsylvania, ...	88 1 0	81 1 0
	Virg'a and Mary'd,	1,367 15 0	713 6 0
1731–2..	Pennsylvania, ...	52 11 0	54 00 0
	Virg'a and Mary'd,	1,504 8 0	721 00 0
1732–3..	Pennsylvania, ...	43 3 0	52 00 0
	Virg'a and Mary'd,	1,669 14 0	639 17 0
1733–4..	Pennsylvania, ...	59 4 0	88 00 0
	Virg'a and Mary'd,	1,397 17 0	644 4 0
1734–5..	Pennsylvania, ...	197 14 0	44 17 0
	Do.	00 00 0	19 17 bar iron.
	Virg'a and Mary'd,	1,353 3 0	1,059 00 pig.
1739....	Pennsylvania, ...	170 5 0	44 9 bar iron.
	Virg'a and Mary'd,	2,242 2 0	00 00
1740....	Pennsylvania, ...	159 00 0	00 00
	Virg'a and Mary'd,	2,020 2 0	5 tons bar iron.
1741....	Pennsylvania, ...	153 00 0	00
	Virg'a and Mary'd,	3,264 1 0	5 do.
1742....	Pennsylvania, ...	143 00 0	00
	Virg'a and Mary'd,	1,926 3 0	00
1743....	Pennsylvania, ...	62 12 0	00
	Virg'a and Mary'd,	2,816 1 0	57 tons do.
1744....	Pennsylvania, ...	87 15 0	85 do.
	Virg'a and Mary'd,	1,748 00 0	00

COLONIES.		LONDON.			OUTPORTS.
		T.	C.	Qr.	
1745....	Pennsylvania, ...	97	7	0	00
	Virg'a and Mary'd,	2,130	16	0	5 tons bar iron.
1746....	Pennsylvania, ...	103	1	0	00
	Virg'a and Mary'd,	1,729	00	0	195 do.
1747....	Pennsylvania, ...	24	14	0	00 do.
	Virg'a and Mary'd,	2,119	00	0	83 do.
1748....	Pennsylvania, ...	114	10	0	00
	Virg'a and Mary'd,	2,017	00	0	00

PIG AND BAR IRON EXPORTED BY THE AMERICAN COLONIES FROM 1750 TO 1756.

COLONIES.		BAR.			PIG.		
		T.	C.	Qr.	T.	C.	Qr.
1750.	Carolina, ...				00	00	0
	New England,....				21	1	12
	New York, ...				75	12	1
	Pennsylvania, ...				318	9	3
	Virg'a and Mary'd,				2,508	16	1
1751.	Carolina, ...				17	4	0
	New England,....				9	14	0
	New York, ...				33	0	3
	Pennsylvania, ...				199	15	0
	Virg'a and Mary'd,				2,950	5	3
1752.	Virg'a and Mary'd,				20	0	0
	Carolina, ...				00	0	0
	New England,....				41	5	0
	New York, ...	64	16	2	156	8	2
	Pennsylvania, ...	16	10	2	2,762	8	0
1753.	Carolina, ...	00	00	0	10	0	6
	New England,....	2	8	0	40	10	0
	New York, ...	00	00	0	97	4	3
	Pennsylvania, ...	107	13	2	242	2	15
	Virg'a and Mary'd,	97	16	0	2,347	9	2
1754.	Carolina, ...	00	00	0	20	0	0
	New England,....	00	00	0	4	16	0
	New York, ...	6	10	0	115	16	2
	Pennsylvania, ...	110	9	3	512	19	3
	Virg'a and Mary'd,	153	15	1	2,591	4	3

	COLONIES.	BAR.			PIG.		
		T.	C.	Qr.	T.	C.	Qr.
1755.	Carolina, ...	00	00	0	14	0	3
	New England,....	00	00	0	00	00	0
	New York, ...	11	12	0	457	8	0
	Pennsylvania, ...	79	5	0	836	6	1
	Virg'a and Mary'd,	299	1	3	2,132	15	1

In 1761 the Governor and Council of Maryland reported to the Commissioners of the Board of Trade and Plantations in England, that there were eighteen furnaces and ten forges in that State, which made 2,500 tons of pig and 600 tons of bar iron.

It was, no doubt, the scarcity of wood, and high price of iron in Great Britain, which stimulated the colonies to manufacture iron; for, up to this period, England had made so little progress that her production of pig iron had only reached 17,000 tons. The seat of iron manufactures were then principally confined to Pennsylvania, New York, New Jersey, Virginia, and Maryland. In 1730 the Reading furnace was built, and the Warwick in 1736. The Cornwall furnace, in Lebanon county, Penn., so famous for having enriched all its late proprietors, was built in 1741.

The price of pig iron at this establishment, in 1780, was three hundred pounds of Continental money; in 1789, £6 10*s*., Pennsylvania currency; in 1796, £10 10*s*. ($28 00); and in 1800, £10 ($26 67⅓) per ton. The rates of wages were less then than at the present day, averaging about fifteen dollars per month. In 1751 a furnace was built on the Sterling Estate in Orange county, N. Y., which produced annually 1,500 tons of pig iron, which was worked up into bar iron. The great iron chain which crossed the Hudson River, during the Revolution, each link of which weighed 140 pounds, was made at these works.

The Mount Etna furnace, near Hagerstown, was among the first in this country to cast cannon. An eighteen-pounder of its manufacture is still exhibited at Barracks Hill, near Fredericktown, as a revolutionary relic.

In the spring of 1765, the Act of Parliament of 1750 was so far modified as to permit American iron to be imported into Ireland free of duty; and from this time forward the American colonies made rapid progress in the manufacture of iron and steel, which returned large profits to the manufacturer.

PIG AND BAR IRON EXPORTED BY THE AMERICAN COLONIES TO GREAT BRITAIN FROM 1760 TO 1776.

YEARS.	BAR IRON.				PIG IRON.			
	Tons.	Cwt.	Qr.	lb.	Tons.	Cwt.	Qr.	lb.
1761	39	1	0	0	2,766	2	3	12
1762	122	12	2	14	1,763	6	0	2
1763	310	9	3	2	2,566	8	0	25
1764	1,059	8	0	10	2,554	8	3	21
1765	1,078	16	0	16	3,264	8	1	22
1766	1,257	14	3	9	2,887	5	1	15
1767	1,325	19	0	18	3,323	2	1	19
1768	1,989	11	0	16	2,953	0	2	14
1769	1,779	13	0	23	3,401	12	2	9
1770	1,716	8	0	21	4,232	18	1	18
1771	2,222	8	1	24	5,303	6	3	13
1772	965	15	0	23	3,724	19	2	11
1773	837	3	3	16	2,937	13	0	23
1774	639	2	0	23	3,451	12	2	14
1775	916	5	2	11	2,996	0	2	16
1776	28	0	1	8	316	1	3	8

In September, 1774, the colonies entered into an agreement not to export to Great Britain any merchandise or manufactures, nor to import from the mother country any British manufactures, until their grievances should be redressed. They also declared that they would "not import or purchase any slaves imported after the first of December, and also wholly *to discontinue the slave-trade*, and neither hire vessels nor sell commodities or manufactures to those who are concerned in it."

On the 4th of July, 1776, the American colonies declared themselves independent, and Congress, while in session, passed a law exempting all persons then engaged in the manufacture of iron from performing military duty.

The war with Great Britain gave a great impetus to all branches of iron manufacture while it continued. In several of the colonies large quantities of charcoal, pig, and bar iron of a superior quality were produced. The Salisbury and Livingston bar iron commanded high prices, and were much esteemed in Great Britain.

The Andover Works in New Jersey manufactured, during the Revolution, both iron and steel for the army.* The iron works and founderies of New England, Maryland, Pennsylvania, and Virginia, also supplied the army with cannon, shell, and shot.

No statistics exist which show the exact production and consumption of iron in the colonies during the

* *Resolved*, That a letter be written by the Board of War to the Governor and Council of the State of New Jersey, setting forth the peculiarity of the demand for their works, *being the only proper means for procuring iron for steel*, an article without which the service must irreparably suffer; and that the said Governor and Council be desired to take such means as they shall think most proper for putting the said works in blast, and obtaining a supply of iron without delay.—*Journals of Congress.*

War of the Revolution, and for some years afterwards. However, it may be safely set down at about thirty thousand tons per annum. In 1783, Mr. Cort obtained in England two patents, one for the puddling and the other for the rolling of iron, which gave a great impulse to its production. The object of these processes was to convert into malleable iron, cast or pig iron, by means of the flame of pit coal, in a common air furnace, and to form the result into bars by the use of rollers in the place of hammers.

On the return of peace, in 1783, the trade and commerce of the United States revived. New markets were thrown open to American enterprise. The commercial relations between Great Britain and the United States were governed by separate and distinct regulations of the several States, on the one side, and general orders in council on the other.

The first order in council issued after the Independence of the United States was declared, placed the exportation of bar and pig iron on the same footing of duties as the like merchandise exported from any of the British possessions.

The old confederation having made no provision to protect home manufactures, Great Britain, true to her policy of monopolizing every market, in the first two years after the peace, flooded the American market with every description of merchandise, which involved the United States in debt, and threatened for a time the peace and existence of the Union.

The manufacture of iron fell off rapidly. Most of the iron works in New England and the middle States were closed by the sheriff, and the exports of bar and pig iron were reduced more than one-half. From

1789 to 1790 the United States exported but 200 tons of bar iron and 3,500 tons of pig iron, so completely had Great Britain taken possession of the market. From this time also may be dated the commencement of her system of protection to her own manufactures and prohibition to all others, by which the domestic competition of forty-one years enabled her to manufacture iron at prices defying all rivalry, and throw open her ports and proclaim free trade when she knew that no nation could compete with her.

STATEMENT OF IRON IMPORTED FROM RUSSIA FROM 1783 TO 1804.

YEAR.	BARS.	NAIL ROD IRON.
	Poods.	Poods.
1783..........	6,615	....
1784..........	6,612	....
1785..........	38,618	....
1786..........	31,858	2,322
1787..........	10,833	1,260
1788..........	17,054	846
1789..........	24,981	1,250
1790..........	78,160	2,526
1791..........	48,136	2,621
1792..........	132,380	1,132
1793..........	177,826	1,071
1794..........	250,635	694
1795..........	206,039	504
1796..........	296,691	6,405
1797..........	112,200	506
1798..........	142,654	1,250
1799..........	239,885	126
1800..........	112,568	314
1801..........	269,709	426
1802..........	309,425	21
1803..........	413,822	....
1804..........	278,264	....

In 1787, a general convention was held in Philadelphia to frame a new Constitution which went into operation on the 4th of March, 1789.

One of the objects which claimed the attention of Congress under the new Constitution, was the protection and encouragement of iron manufactures, which had been much depressed by the free-trade policy adopted by the Federal Government.

In 1790, Congress directed the Secretary of the Treasury (Hamilton) to make a report on the subject of protecting home manufactures.

In this report he particularly recommended the manufacture of iron to the special protection of Congress. "The manufacture of this article," he says, "is entitled to a preëminent rank. None is more essential nor extensive in its use." He proposed, that, in order to encourage the production and manufacture of iron, a further duty be laid upon all foreign iron imported into the United States, as well as upon fire-arms, and military weapons. This recommendation appeared to be the more necessary at that time, because Great Britain had, but a few years before, in order, if possible, to prevent foreign competition, passed an act prohibiting the carrying out of the kingdom all the various tools and instruments, models and plans, necessary to be used in the manufactures of iron or military weapons under a heavy fine and imprisonment.

The protection thus proposed by Congress created no little solicitude on the part of Great Britain, and she agreed, without further hesitation, to negotiate a commercial treaty with the United States, which was signed on the 19th November, 1794. By this treaty, no higher duties were to be paid by either than were paid by all other nations.

The breaking out of the Indian wars soon after, and the renewal of the militia system of the United States, greatly revived the manufacture of fire-arms, notwithstanding the restrictive steps which had been taken by Great Britain to prevent the introduction of suitable machinery to carry on this branch of manufactures in this country.

In 1792, Congress imposed a duty of $20 per ton on steel, and $36 per ton on iron cables imported from Great Britain.

The duties laid by Congress for the purpose of revenue, to take effect after the first of July, 1794, if imported in American vessels, were fixed at 15 per cent. on steel and rolled iron; 10 per cent. on hardware, and 15 per cent., *ad valorem*, on all other manufactures of iron, steel, or brass, and 10 per cent. additional if imported in foreign vessels.

The prices of bar iron, in this country, from 1793 to 1807, were as follows:

From 1793 to 1797	$90 00 to $95 00	per ton.
" 1800 to 1801	100 00 to 105 00	"
" 1803 to 1807	110 00 to 115 00	"

The first considerable rise in foreign iron took place in 1796, when it suddenly advanced 30 per cent.,*

* This was the year before the bank restriction, and the rise occurred in consequence of the importation having fallen off, instead of keeping pace with the increasing demand for consumption in this country, and in the rest of Europe as well as in the United States. Between 1796 and the close of 1800, there was no further advance; but the embargo in Russia, in the latter year, had the effect of raising the price. The advance altogether, including the new duty of £1 per ton, was nearly £10 per ton, and this operated as a sufficient premium for applying increased capital to the production of iron, and bringing into operation

which had the effect to stimulate the manufacture of it in the United States. In Great Britain it operated as a premium in the employment of additional capital in its manufacture, which, together with improved machinery and new methods of making it, greatly increased the production of that country and furnished a large surplus for exportation.

The imports of iron from Sweden and Norway, from 1795 to 1801, averaged about $80,000 per annum. In 1802, 1803, and 1804, the United States imported annually from Great Britain iron manufactures valued at upwards of one million and a half of dollars, and for the next three years the imports greatly exceeded all former years. The manufacture of iron in the United States during this period could scarcely be maintained, and must have perished if the various non-importation, non-intercourse acts, the embargo, and war of 1812, had not come to its relief.

The furnaces, up to this time, were blown by wooden and leather bellows, and the cold blast, and one tuyère, and their working profitably was greatly influenced by the skill and influence of the founder. With the greatest skill and every precaution, the yield still depended much upon the blast, and as this was invariably produced by water-power, which was often very irregular and weak, and in dry seasons incapable of furnishing the necessary power, it will account for the small average *annual* product of the furnaces of that day. The ores used were mainly the

all the powers of machinery. Thenceforward the production of iron proceeded so rapidly, that with the aid of further duties, amounting almost to a prohibition of importation, it not only kept pace with the increasing demand, but nearly superseded the use of foreign iron in England, and furnished a surplus for exportation.—TOOKE.

hematites and magnetic ores, and the weekly average of the furnaces was from twenty to twenty-five tons per week.

MATERIALS.	QUANTITIES.	COST.
Charcoal,	240 lbs. at 3 cents,	$7 20
Ore,	3 tons at $2 00,	6 00
Limestone,		30
Labor,		4 00
Incidental—wear and tear, etc.,		2 50
Capital, estimated upon an investment of $50,000 for real estate business, etc., and on a production of 800 tons to the furnace, per annum,		3 75
		23 75
Profit for deterioration of estate, etc.,		5 00
Total cost,		$28 75

The pig metal was sold to forges in the neighborhood, for but few were wealthy enough to carry it throughout the entire process. The cost of transportation varied from $3 to $5 per ton. It was then converted into bar iron, without refining.

The pig metal, by the time it reached the forge, cost the iron master $33 75 per ton; and to convert it into bar iron, $83 25 per ton.

MATERIALS.	QUANTITIES.	COST.
Pig metal,	30 cwt. at $34 50,	$51 75
Charcoal,	375 lbs. at 3 cents,	11 25
Labor,		13 25
Incidental expenses,		2 50
Interest on capital—say $15,000, and producing 200 tons of bar,		4 50
Total cost,		$83 25

The great expense of making iron at that time arose not only from the expense of cutting, charring, and hauling wood, and transporting the same to the furnace, but afterwards the crude iron to the forge, and from the forge to the rolling-mill, for the purpose of manufacturing into merchant bar, and then again to the slitting and nailing factory, and all these establishments were sometimes many miles from each other; and after this hauling and transportation, carried at an enormous cost, in many cases, to the Atlantic market, where it would probably command from $110 to $120 per ton, and blooms from $75 to $80 per ton.

In 1810, the Secretary of the Treasury presented a report to Congress on the manufacture of iron, which he represented as being now firmly established, the value of which was estimated at from fifteen to twenty millions of dollars. By this report, founded on official documents, there were in the United States 153 furnaces, which made 53,908 tons of pig iron, and 330 forges, which made 24,541 tons of bar iron; 316 trip hammers, and 34 rolling and slitting mills, which required 6,500 tons of iron; and 410 naileries, which made 15,727,914 lbs. At this period, iron was exclusively made with charcoal—from the smelting of the ore in the blast furnace to the finished bar in the forge fire. From 1810 to 1812, the capital and enterprise of the country was too lucratively employed in common to be much attracted to the iron trade, until the war with Great Britain was declared, when the necessity of the country for this indispensable article advanced the price enormously, and encouraged the erection of extensive works which enriched all those who engaged in its manufacture.

STATEMENT SHOWING THE POPULATION AND IRON MANUFACTURES OF THE UNITED STATES AND TERRITORIES FOR THE YEAR 1810.

States.	Population.	Manufactures of Pig Iron and Castings.	Manufactures of Wrought Iron.
Columbia, District of	24,023	....	.....
Connecticut........	264,042	$46,180	$351,198
Delaware..........	72,674	.....	195,420
Georgia	252,433	.....	30,155
Kentucky	406,511	1,000	44,260
Louisiana	76,556	.....	244,000
Maine, District of ..	228,705	.....	21,929
Maryland	380,546	249,653	491,058
Massachusetts	472,040	154,700	2,078,542
New Hampshire....	214,360	.....	170,350
New Jersey	245,555	861,932	526,511
New York	959,049	362,020	497,875
North Carolina.....	555,500	135,160	554,950
Ohio	230,760	109,090	74,123
Pennsylvania	810,091	1,301,343	4,492,478
Rhode Island......	77,031	3,970	56,770
South Carolina.....	415,115	.....	90,227
Tennessee	261,727	98,097	263,327
Vermont	217,713	122,000	222,059
Virginia	974,622	171,312	538,854
Territories. Illinois	12,282	.....	.....
Territories. Indiana	24,520	.....	4,000
Territories. Michigan......	4,762	.....	.....
Territories. Mississippi	40,352	.....	.....
Territories. Missouri......	20,845	.....	.....
Total	7,239,814	3,616,457	10,998,086

In 1812, Congress laid an embargo on the commerce of the United States, which was followed by a declaration of war against Great Britain. In order to meet the expenses of war, Congress, on the 30th June, increased the rate of duties one hundred per cent. on all imports. Iron manufactures rapidly increased under this tariff, until the return of peace, when the Government once more adopted the free-trade policy, which ruined most of the iron-masters, and prostrated for some years this important branch of industry. The production of pig iron fell off from 54,000 tons to about 20,000 tons annually for several years.

In 1816, Congress passed a new tariff. The duty on bar iron, rolled, was fixed at $30 per ton ; not rolled, $9 per ton ; and steel, $20 per ton. It did not, however, prevent Great Britain from flooding the States with her manufactures. The total value of all imports, from Oct. 31, 1815, to Sept. 30, 1816, amounted to the enormous sum of $155,302,706, and for the three fiscal years ending Sept. 30, 1815, 1816, and 1817, to $359,394,274.

ENGLISH IRON, UNWROUGHT STEEL, HARDWARE, AND CUTLERY, IMPORTED INTO THE UNITED STATES FROM 1815 TO 1839.

Year.	Tons.	Cwt.	Year.	Tons.	Cwt.
1815........	21,501	15	1828........	22,865	10
1816........	21,634	4	1829........	17,387	10
1817........	10,725	3	1830........	21,330	5
1818........	13,737	10	1831........	41,452	6
1819........	8,251	3	1832........	45,436	15
1820........	8,199	0	1833........	62,253	4
1821........	9,562	3	1834........	47,676	1
1822........	15,835	9	1835........	63.012	10
1823........	13,841	11	1836........	91,387	14
1824........	11,781	14	1837........	54,120	2
1825........	13,037	4	1838........	78,039	17
1826........	12,491	2	1839........	85,171	11
1827........	21,855	11			

American statesmen now fully realized the folly and the injurious effects of the free-trade system upon the home industry of the country, and in 1818 Congress raised the duty on foreign iron, which revived the manufacture of it in this country for a few years. The duty on hammered iron was fixed at $15 per ton, pig iron at $10, and rolled iron at $30 per ton.

In consequence of the heavy cost of transportation of iron to Pittsburg, in the years 1818 and 1819, English bar iron sold in that city from $190 to $200 per ton; boiler iron, $350 per ton; sheet iron, $360 per ton; and hoop iron at $250 per ton. At Cincinnati, the price of bar iron from 1814 to 1818 averaged $200 per ton, and castings, $120 per ton; while hammered iron sold in the Atlantic ports of the United States at the following prices:

1814	From $125	to	$148	per ton.
1815	" 130	to	150	"
1816	" 110	to	120	"
1817 to 1819	" 90	to	100	"
1821 to 1822	" 85	to	95	"
1823 to 1824	" 90	to	95	"

Although no efforts had been made to manufacture iron until after the tariff of 1818 was passed, yet some improvements had been made in its manufacture by reserving the charcoal for the blast furnace, and rolling the iron from the bloom, with bituminous coal, into boiler, plate, sheet iron, hoops, and all the variety of the higher descriptions of iron.

The delicate process of puddling was not yet introduced, although it had been thoroughly and profitably carried out in Great Britain, some years before, because the internal improvements, canals and railways, had not yet reached those rich deposits of mineral wealth which, like those of England, lay far in the interior; but if this process had been thoroughly understood, iron could not have been manufactured at lower rates, upon the Atlantic seaboard, than that manufactured with charcoal.

It was therefore necessary, in order to adopt some of the English improvements, that canals and railways should be constructed, so as to extend to the manufacturer cheap transportation to a market.

The Lehigh navigation penetrated the coal region of Mauch Chunk, in 1820, and the Schuylkill canal in 1825, which enabled the manufacturer to purchase coal, in the latter year, for half the price that foreign coal sold at in the Atlantic ports.

It was not until after these improvements had been made, that the process of puddling was introduced into Pennsylvania, which gave so great an impetus to her iron trade; and Pittsburg, located in the richest of the bituminous coal fields of this country, soon rendered the western States independent of foreign markets, and became, from the skill of her manufacturers, the Birmingham of the United States. In a few years, they reduced the price of iron *one-half* to the consumer of that region, notwithstanding they had to transport their blooms for the manufacture of it across the lofty Alleghanies.

STATEMENT SHOWING THE POPULATION AND IRON MANUFACTURES OF THE UNITED STATES AND TERRITORIES FOR THE YEAR 1820.

States.	Population.	Manufactures of Pig Iron and Castings.	Manufactures of Wrought Iron.
Alabama	127,901	.....	$15,620
Columbia, District of .	33,039	.....	5,000
Connecticut.........	275,202	.....	296,260
Delaware...........	72,749	$30,000	30,000
Georgia............	340,987	.....	69,036
Indiana	147,178	.....	3,000
Kentucky	564,317	130,000	138,800
Louisiana	153,407	.....	10,000
Maine	298,335	.....	65,200
Maryland	407,350	93,000	449,080
Massachusetts.......	523,287	77,500	423,610
Mississippi..........	75,448	.....	.....
Missouri	66,586	.....	18.421
New Hampshire	244,161	40,500	18,340
New Jersey.........	277,575	76,300	188,997
New York..........	1,372,812	342,400	472,158
North Carolina	638,829	.....	53,510
Ohio...............	581,434	413,350	491,707
Pennsylvania........	1,049,458	563,810	1,156,266
Rhode Island	83,059	.....	19,032
South Carolina......	502,741	.....	42,000
Tennessee	422,813	184,916	246,755
Vermont	235,764	85,400	33,340
Virginia	1,065,379	193,100	393,417
Arkansas Territory...	14,273	.....	120
Illinois Territory	55,211	.....	.....
Michigan Territory...	8,896	.....	1,000
Total............	9,638,131	2,230,275	4,640,669

In 1823, President Monroe recommended to Congress a revision of the existing tariff.

It was not, however, until 1824, when the results of the extravagant legislation of Great Britain upon the subject of her Corn Laws had been felt in this country, in its full extent, that the agricultural interests of the middle States, and those of the eastern and western States, were compelled to unite with the manufacturing interests in petitioning Congress to impose a higher rate of duties on foreign manufactures. The tariff on foreign iron was again modified, but it proved a very imperfect measure of protection against the combination of cheap capital and labor of England.*

The imports of foreign iron for the ten years previous to 1826 averaged about ten per cent. of the entire exports of Great Britain.

Under the treaty negotiated with Sweden, July 4,

* The disposition of England to bring against the rising industry of foreign countries the full power of capital and competition, is openly avowed. In a recent report made to the British Parliament by the Commissioner appointed under the provisions of the Act 5 and 6 Victoria, c. 99, to inquire into the operation of that Act, and into the state of the population in the mining districts, 1854, the following passage is found:

"I believe that the laboring classes generally, in the manufacturing districts of this country, and especially in the iron and coal districts, are very little aware of the extent to which they are often indebted for their being employed at all, to the immense losses which their employers voluntarily incur in bad times, in order to destroy foreign competition and to gain and keep possession of foreign markets. Authentic instances are well known of employers having at such times carried on their works at a loss amounting in the aggregate to three or four hundred thousand pounds, in the course of a few years. If the efforts of those who encourage the combinations to restrict the amount of labor, and to produce strikes, were to be successful for any length of time, the great accumulations of capital could no longer be made which enable a few of the most wealthy capitalists to overwhelm all foreign competition in times of great depression, and thus to clear the way for the whole trade to step in when prices revive, and to carry on a great business before foreign capital can again accumulate to such an extent as to be able to establish a competition in prices with any chance of success.

"The large capitals of the country are the great instruments of warfare (if the expression may be allowed) against the competing capitals of foreign countries, and are the most essential instruments now remaining by which our manufacturing supremacy can be maintained; the other elements—cheap labor, abundance of raw materials, means of communication, and skilled labor—being rapidly in process of being equalized."

1827, the trade with that country was based upon the liberal principles of entire reciprocity.

STATEMENT OF BAR IRON IMPORTED FROM SWEDEN FROM 1830 TO 1838.

BAR IRON.		BAR IRON.	
Years.	Tons.	Years.	Tons.
1830	15,532	1834	19,638
1831	23,133	1835	28,728
1832	20,122	1836	27,342
1833	19,100	1838	25,669

BAR IRON (INCLUDING CAST AND OTHER STEEL) IMPORTED FROM SWEDEN AND NORWAY FROM 1845 TO 1855.

Years.	Cwt.	Values.
1845	272,496	$626,166
1846	256,663	717,116
1847	288,464	609,729
1848	290,082	740,078
1849	295,359	729,206
1850	397,231	1,025,587
1851	409,003	942,961
1852	289,391	773,674
1853	181,049	445,808
1854	203,137	510,221
1855	296,500	844,233

In 1828, Mr. Neilson, of Glasgow, took out a patent for the application of hot-blast in the manufacture of iron, which was economical in its result, and was generally adopted in the United States, and greatly increased the production of iron. To this great discovery, the development of the railway system of this country and Europe is, perhaps, more indebted than any other.

In 1828, Congress passed a new tariff, and increased

the rate of duties on iron; but it still proved insufficient to prevent foreign competition.

Pig iron, ...	$12 50 per ton.
Hammered iron, . . .	22 40 "
Rolled iron, ...	37 50 "

The receipts of the Treasury, under this tariff, gradually improved, and soon enabled the Government to pay off her war debt.

In some parts of the Union, the tariff of 1828 was looked upon as an unequal and oppressive measure, and South Carolina in particular took steps, on the ground that it was unconstitutional to impose taxes to protect particular interests, to resist its execution.

In consequence of that opposition, it was slightly modified in 1832; but she did not cease to complain until Congress was compelled, in 1833, to reduce the rates still further, by providing for a gradual reduction on imports.

By this act, everything was sacrificed to the speculative theories of the day, and at a time when the *extension* of this industry, and the improvement, economy and skill introduced into it, had reduced the price of iron to a lower point than it had ever reached before in this country.

The duty on English bar iron was gradually reduced from $30 per ton, in 1832, to $7 50 in July and August, 1842; and pig metal from $9 50 per ton, in 1834, to $4 50 per ton in 1842. Railway iron was admitted free of duty. The effect of this great reduction of duties on the importation of iron may be seen in the following tables:

STATEMENT OF IRON AND STEEL IMPORTED INTO THE UNITED STATES, FROM 1828–29 TO 1831–32.

ARTICLES.	1828–29.		1829–30.		1830–31.		1831–32.	
	Tons.	Export value.	Tons.	Export value.	Tons.	Export value.	Tons.	Export value.
Bar and bolt iron, rolled, ...	3,320	$119,326	6,449	$226,336	17,245	$544,664	20,387	$701,549
Bar and bolt iron hammered, or otherwise manufactured,	29,489	1,884,069	30,693	1,730,375	23,308	1,260,166	38,150	1,929,493
Pig iron, ...	1,138	28,811	1,129	25,664	6,448	160,681	10,151	222,303
Hoop and sheet iron, ...	1,089	89,057	1,038	59,822	2,532	151,900	2,853	182,559
Brazier's rods, 3–16 a 8–16, inclusive, ...	75	6,164	97	5,945	217	13,660	233	13,727
Nail and spike rods, slit, ...	3	234	14	784	101	4,585	56	2,063
Band, scroll, or casement rods, slit or hammered, ...	—	—	$1\frac{1}{4}$	81	10	72	3	176
Old or scrap iron, ...	—	—	—	—	—	—	—	—
Total iron, ...	35,114	2,127,661	39,421	2,049,007	49,861	2,135,728	71,833	3,051,870
Steel, ...	1,200	289,931	1,223	291,957	1,710	399,635	2,146	645,510
Total iron and steel, ...	36,314	$2,417,592	40,644	$2,340,964	51,571	$2,535,363	73,979	$3,697,380

STATEMENT OF IRON AND STEEL IMPORTED INTO THE UNITED STATES, FROM 1832–33 TO 1835–36.

ARTICLES.	1832–33.		1833–34.		1834–35.		1835–36.	
	Tons.	Export value.	Tons.	Export value.	Tons.	Export value.	Tons.	Export value.
Bar and bolt iron, rolled, ...	28,028	$1,002,750	28,896	$1,187,236	28,410	$1,050,152	46,675	$2,131,828
Bar and bolt iron, hammered, or otherwise manufactured,	36,124	1,837,473	31,784	1,742,883	31,524	1,641,359	32,987	1,891,214
Pig iron, ...	9,330	217,668	11,113	270,325	12,295	289,779	8,541	272,978
Hoop and sheet iron, ...	3,350	245,848	2,214	190,237	2,009	133,639	3,643	325,676
Brazier's rods, 3–16 a 8–16, inclusive, ...	221	12,834	132	10,017	113	7,428	240	21,764
Nail and spike rods, slit, ...	95	6,080	$\frac{3}{4}$	77	$1\frac{1}{4}$	244	10	1,301
Band, scroll, or casement rods, slit or hammered, ...	12	2,063	3	230	$\frac{1}{20}$	5	$\frac{1}{8}$	5
Old or scrap iron, ...	998	24,035	1,617	33,243	640	10,609	1,846	28,224
Total iron, ...	78,158	3,348,751	75,759	3,434,248	74,992	3,133,215	93,342	4,672,990
Steel, ...	2,131	523,116	2,431	554,150	2,605	576,888	2,878	686,141
Total iron and steel, ...	80,289	$3,871,867	78,190	$3,988,398	77,597	$3,710,103	96,220	$5,359,131

STATEMENT OF IRON AND STEEL IMPORTED INTO THE UNITED STATES, FROM 1836–37 TO 1839–40.

ARTICLES.	1836–37.		1837–38.		1838–39.		1839–40.	
	Tons.	Export value.	Tons.	Export value.	Tons.	Export value.	Tons.	Export value.
Bar and bolt iron, rolled, ...	47,839	$2,573,367	36,174	$1,825,121	60,285	$3,181,180	32,825	$1,707,650
Bar and bolt iron, hammered, or otherwise manufactured,	31,325	2,017,346	21,319	1,166,196	35,557	2,054,094	28,819	1,689,831
Pig iron, ...	14,128	422,929	12,192	319,099	12,507	285,300	5,516	114,562
Hoop and sheet iron, ...	5,041	504,473	2,536	218,192	3,309	354,933	2,469	235,809
Brazier's rods, 3–16 a 8–16, inclusive, ...	201	21,792	142	10,648	381	27,942	193	47,782
Nail and spike rods, slit, ...	½	32	1⅓	94	36	2,291	½	24
Band, scroll, or casement rods, slit or hammered, ...	¼	36	55	2,712	15	886	15	963
Old or scrap iron, ...	766	18,391	436	7,567	589	10,161	707	15,749
Total iron, ...	99,300	5,558,366	72,855	3,549,629	112,679	5,916,787	70,544	3,812,370
Steel, ...	3,566	804,817	1,907	487,334	2,958	771,809	2,225	528,716
Total iron and steel, ...	102,866	$6,363,183	74,762	$4,036,963	115,637	$6,688,596	72,769	$4,341,086

STATEMENT OF IRON AND STEEL IMPORTED INTO THE UNITED STATES, FROM 1840–41 TO 1843–44.

ARTICLES.	1840–41.		1841–42.		1842–43.		1843–44.*	
	Tons.	Export value.	Tons.	Export value.	Tons.	Export value.	Tons.	Export value.
Bar and bolt iron, rolled, ...	63,055	$2,172,278	61,600	$2,053,453	20,230	$637,617	46,000	$1,825,121
Bar and bolt iron, hammered, or otherwise manufactured,	29,605	1,614,420	19,512	1,041,410	8,440	450,317	17,500	855,220
Pig iron, ...	12,267	223,288	18,694	295,284	6,472	76,858	26,050	349,600
Hoop and sheet iron, ...	3,646	376,075	3,560	296,679	1,522	154,638	3,600	280,360
Brazier's rods, 3–16 a 8–16, inclusive, ...	164	12,843	530	37,767	212	15,369	470	10,648
Nail and spike rods, slit, ...	13½	613	18	860	10	730	27	1,890
Band, scroll, or casement rods, slit or hammered, ...	15	1,161	22	1,023	10	1,612	60	6,500
Old or scrap iron, ...	783	10,537	685	8,207	169	4,424	5,770	152,160
Total iron, ...	109,548	4,411,215	104,621	3,734,683	37,071	1,341,565	99,477	3,481,499
Steel, ...	2,563	609,201	2,771	597,317	1,334	324,086	2,800	487,334
Total iron and steel, ...	112,111	$5,020,416	107,392	$4,332,000	38,405	$1,665,651	102,277	$3,968,833

* The last quarter of 1844 is only estimated in part.

And as this was only the beginning of the experimental system, which was carried out with such disastrous results to the industry of the country, and to the working classes of society, it will be well to examine how far it realized, for the consumers, *low prices.*

				£	s.	d.			£	s.	d.
In 1832	bar iron	sold in	England for	5	0	0,	Railway	iron,	6	15	0
1833	"	"	"	6	0	0	"	"	7	10	0
1834	"	"	"	6	10	0	"	"	8	00	0
1835	"	"	"	5	15	0	"	"	8	5	0
1836	"	"	"	10	00	0	"	"	11	15	0
1837	"	"	"	8	15	0	"	"	10	00	0
1838	"	"	"	8	15	0	"	"	10	10	0
1839	"	"	"	9	00	0	"	"	10	10	0
1840	"	"	"	8	00	0	"	"	9	12	6
1841	"	"	"	6	10	0	"	"	8	00	0

Thus teaching the country again, that the only effect of low duties was to prostrate the iron trade, enrich the foreign manufacturer and laborer, and bring their establishments into the highest perfection and improvement, without reducing the cost to the consumer in the United States.

STATEMENT SHOWING THE POPULATION AND IRON MANUFACTURES OF THE UNITED STATES AND TERRITORIES FOR THE YEAR 1830.

STATES.	Census for 1830.	Manufactures of Pig Iron and Castings.	Manufactures of Wrought Iron.
Alabama..........	309,527	—	—
Columbia, District of	39,834	—	—
Connecticut	297,675	$136,762	$500,000
Delaware	76,748	—	160,000
Georgia	516,823	—	—
Illinois	157,445	—	—
Indiana...........	343,031	—	—
Kentucky	687,917	—	—
Louisiana	215,739	—	—
Maine	399,455	54,500	608,500
Maryland	447,040	—	—
Massachusetts......	610,408	1,437,147	8,360,102
Mississippi	136,621	—	—
Missouri	140,455	—	—
New Hampshire....	269,328	52,891	364,284
New Jersey	320,823	412,941	642,238
New York	1,918,608	751,807	1,989,790
North Carolina	737,987	—	—
Ohio	937,903	—	—
Pennsylvania	1,348,233	1,643,702	3,762,847
Rhode Island	97,199	139,973	200,000
South Carolina.....	581,185	—	—
Tennessee.........	681,904	—	—
Vermont..........	280,652	127,680	149,490
Virginia	1,211,405	—	—
Territories. Arkansas	30,388	—	—
Territories. Florida	34,730	—	—
Territories. Michigan......	31,639	—	—
Territories. Naval service..	5,318	—	—
Total..........	12,866,020	4,757,403	16,737,251

A committee appointed by a convention of manufacturers of iron in Philadelphia, in 1831, reported the amount of iron made in the United States to be as follows :

	1828.	1830.
Pig iron, ...	108,564 tons.	137,075 tons.
Castings from the ore at blast furnaces, ...	14,840 "	18,273 "
Bloomed bar iron made from the ore, equal in pig iron, at 28 cwt. pig iron to a ton of bar iron, at	7,477 "	8,194 "
Total made, reckoned in pigs and castings, ...	130,881 tons.	163,542 tons.

The pig iron was all converted into castings, bar iron, nails, and other wrought iron.

REPORT ON IRON OF THE CONVENTION OF THE FRIENDS OF DOMESTIC INDUSTRY, HELD IN THE CITY OF NEW YORK, NOVEMBER, 1831.

"The Committee on Iron respectfully report: That, in discharge of the duties assigned them, they have availed themselves of the information obtained by the Convention of manufacturers of iron, recently assembled in Philadelphia (of which several of your Committee were members), which information was originally collected for the purpose of answering the call made upon the Secretary of the Treasury at the close of the last session of Congress; and is, they have every reason to believe, as precise and accurate in all its parts, as any body of facts, of equal magnitude and importance, which, under similar circumstances, has ever been submitted to the public. From abstracts of statements made to that meeting, it appears that at 202 furnaces, known to have been in operation, there were made, in the year 1830, 155,348 tons of iron.

"This iron, further investigation enables your Committee to say, is converted into 90,768 tons of bar iron, and 18,273 tons of castings; which, with the bar iron made at the bloomeries, amounting in that year to 5,858 tons, and making a total of bar iron of 96,621 tons, will, if estimated at the average wholesale prices of the principal markets of the country, give an aggregate value for the production of that year of 11,444,410 dollars.

"The same statements exhibit for the three years ending with 1830 (when the bar iron made at the bloomeries is reduced into pig iron and added to that made at the furnaces) the following results:

Years.	Iron.	Value.
For 1828, . . .	130,881 tons,	10,861,440 dollars.
" 1829, . . .	142,870 "	11,528,134 "
" 1830, . . .	163,542 "	11,444,410 "

Increase in quantity very nearly 25 per cent. Increase in market value not quite 5½ per cent.

"It will be perceived by the last statement, that the increase in value does not keep pace with that of quantity; and your Committee would here, for a moment, call the attention of the Convention to this fact, affording a practical refutation of the doctrine, that an increased impost necessarily enhances the price to the consumer.

"In this instance, the average price of bar iron in 1828 was 118⅓ dollars. In that year, an addition to the duty on hammered iron was made of $4 40 per ton, and on rolled of $7 00. In the following year, the price fell to 114⅔ dollars, and in 1830 to 96⅔ dollars; showing a decline in two years of 21⅔ dollars per ton, in the face of the increased duty above mentioned: a decline effected exclusively by domestic competition, inasmuch as no corresponding diminution of price took place abroad, and the fall here was greatest in those markets which are inaccessible to foreign iron.

"In making these statements, your Committee have been careful to found them upon data which they believe will bear the test of the most rigid scrutiny. They have been particularly cautious to guard against exaggeration. They believe the cause they desire to sustain needs no aid beyond the simple truth.

"In further illustration of the facts above stated, and of other beneficial consequences flowing from the system of protection, they beg leave now to refer to the accompanying statements marked A and B.* By

* Statement A.—*Prices of Iron in England, taken from the Invoices of the Importers in New York.*

YEARS.	FLAT.	ROUNDS.					SQUARE.		
	Common size.	Common size.	$\frac{5}{8}$	$\frac{1}{2}$	$\frac{3}{8}$	$\frac{1}{4}$	Common.	$\frac{1}{2}$	
	£ s. d.	£ s. d.	£ s. d.	£ s. d.	£ s.	£ s.	£ s. d	£ s. d.	£ s. d.
1806	20 6 0	22 6 0	23 6 0	27 0 0			19 6 0		
1808		14 5 0	15 5 0						
1809	14 0 0	15 0 0	16 0 0				14 0 0	17 0 0	
1810	13 0 0	14 0 0		17 0 0			13 0 0	16 0 0	
1815	11 0 0	11 0 0	11 0 0	16 0 0			11 0 0	13 0 0	
1816	10 0 0	10 0 0	10 0 0				10 0 0	12 0 0	
1819	12 6 0	12 6 0	12 6 0				12 6 0		
1820	9 16 9	10 0 0	10 6 0	11 6 0	12 5			11 6 0	
1821	8 15 6	8 15 6	8 15 6	9 5 6	11 14	13 13	8 15 6	9 15 6	
1822	8 0 0	8 0 0	8 0 0	10 3 0			8 0 0		
1823	8 4 0	8 4 6	8 4 6	9 4 6			8 4 0	9 4 0	11 4 0
1824	9 2 0	9 2 0	9 2 0		11 6		9 2 0		
1825	13 15 0	13 15 0	14 6 0				14 6 0	15 6 0	
1826	10 6 0	10 6 0	11 6 0	12 3 0	14 3		10 6 0		
1827	9 13 0	9 13 0	10 0 0		11 7	13 7	9 13 0	10 7 0	11 7 0
Feb., 1828	8 9 0	8 9 0	8 9 0			13 0	8 9 0	9 10 0	11 0 0
July, 1828	7 9 0	7 9 0	7 9 0				7 9 0		10 5 0
1829	6 0 0	6 0 0	6 0 0				6 0 0		
1830	6 0 0	6 0 0	6 0 0	7 0 0	9 0	11 0	6 0 0		
1831	5 10 0 to £6	Same	Same	6 10 0 to £7	9 0	11 0	£5 10 to £6.		

The above prices are pounds, shillings, and pence, sterling, per ton.

In 1806, the difference between the common sizes and half-inch, was nearly the one it will be seen that while iron in some foreign markets advanced from 40 to 50 per cent. from 1824 to 1825, and from 1822 to 1825 experienced fluctuations amounting to nearly 75 per cent. on the lowest

£7 sterling per ton. It required twenty-five years to bring it down to the present difference of about one pound.

Statement of Prices of Iron in Sweden, from 1815 *to* 1831.

Date		£	s.	d.
March, 1815	£ sterling	12	0	0
July, 1816		13	10	0
October, 1816		12	10	0
March, 1817		13	10	0
June, 1817		14	0	0
February, 1819		16	10	0
December, 1819		13	3	0
January, 1820		14	10	0
June, 1821		13	5	0
September, 1821		11	14	0
November, 1822		11	10	0
April, 1823		12	4	0
August, 1823		11	10	0
December, 1823		11	0	0
March, 1824		11	5	0
August, 1824		10	11	0
September, 1824	£ sterling	10	7	6
December, 1824		11	5	0
April, 1825		14	3	0
September, 1825		14	19	0
June, 1826		12	0	0
July, 1827		12	9	0
October, 1827		13	5	0
December, 1827		13	5	0
September, 1828		13	15	0
May, 1829		13	15	0
June, 1829		13	9	0
September, 1829		12	19	0
December, 1829		12	0	0
April, 1830		11	0	0
May, 1831		10	10	0

Wholesale Prices of Hammered Bar Iron in the Seaports of the United States.

1793	90 to 95	dollars	per ton.
1794	" "	"	"
1795	" "	"	"
1796	" "	"	"
1797	100 to 105	"	"
1798	" "	"	"
1799	95 to 100	"	"
1800	" "	"	"
1801	110 to 120	"	"
1802	105 to 110	"	"
1803	" "	"	"
1804	" "	"	"
1805	" "	"	"
1806	" "	"	"
1807	110 to 115	"	"
1808	" "	"	"
1809	" "	"	"
1810	115 to 120	"	"
1811	110 to 115	"	"
1812	" "	"	"
1813	115 to 125	dollars	per ton.
1814	125 to 145	"	"
1815	130 to 150	"	"
1816	110 to 120	"	"
1817	90 to 100	"	"
1818	" "	"	"
1819	" "	"	"
1820	" "	"	"
1821	85 to 95	"	"
1822	" "	"	"
1823	90	"	"
1824	"	"	"
1825	105	"	"
1826	"	"	"
1827	100	"	"
1828	105	"	"
1829	100	"	"
1830	90	"	"
1831	75 to 85	"	"

STATEMENT B.—Showing the effects of a tariff of protection on the article of iron at Pittsburgh and Cincinnati:

In the years 1818, '19, and '20, bar iron in Pittsburgh sold at from 190 to 200 dollars per ton. Now the price is 100 dollars per ton.

cost, our own varied but about 17 per cent., including an additional duty of about 5 per cent. ; and actually receded at a subsequent period, although sustained by a second addition to the duty, to prices below what

In the same year, boiler iron was 350 dollars per ton. Now at 140 dollars per ton.

Sheet iron was but little made in those years, and sold for 18 dollars per cwt. Now made in abundance, and sold at 8½ dollars per cwt.

Hoop iron, under same circumstances, was then 250 dollars, and is now $120.

Axes were then 24 dollars per dozen, and are now 12 dollars.

Scythes are now 50 per cent. lower than they were then—as are spades and shovels.

Iron hoes were in those years 9 dollars per dozen. Now a very superior article of *steel* hoes at 4 to 4½ dollars.

Socket shovels are made at 4½ dollars by the same individual who, a few years ago, sold them at 12 dollars per dozen.

Slater's patent stoves, imported from England, sold in Pittsburgh at 350 to 400 dollars. A much superior article is now made there and sold for 125 to 150 dollars.

English vices then sold for 20 to 22½ cents per lb.; now a superior article is sold at 10 to 10½.

Brazier's rod in 1824 were imported, and cost 14 cents per lb., or $313 60 per ton. Now supplied to any amount of ¼ to ⅜ diameter, at $130 per ton.

Steam engines have fallen in price, since 1823, one-half, and they have one-half more work on them.

The engine at the Union Rolling Mill (Pittsburgh), in 1819, cost $11,000—a much superior one of 130 horse power, for Sligo Mill, cost, in 1826, $3,000.

In 1830, there were made in Pittsburgh one hundred steam engines.

In 1831, one hundred and fifty will be made, averaging $2,000 ; or $300,000 in that article alone.

A two horse power engine costs 250 dollars ; six horse, 500 dollars ; eight to nine horse, 700 dollars. These last are the prices delivered and put up.

At least 600 tons of iron made in Pittsburgh are manufactured into other articles before it leaves the city, from steam engines, of the largest size, down to a threepenny nail.

Eight rolling and slitting mills of the largest power are in the city of Pittsburgh; five of which have been erected since 1828.

Thirty-eight new furnaces have been erected since 1824 in the western parts of Pennsylvania, and that part of Kentucky bordering on the Ohio River, most of them since 1828. The quantity of iron rolled at Pittsburgh was, in

1828	tons,	3,291	19	0	0
1829		6,217	17	0	0
1830		9,282	2	0	0

Being an increase of nearly 200 per cent., in two years.

had prevailed for ten years before, when the existing duty upon hammered iron was but nine dollars per ton, or less than one-half of that now levied.

"This comparative stability, so important to the success of all well-regulated industry, was due exclusively to the domestic supply, which effectually protected the consumer from the foreign speculator, who could otherwise have controlled this market, and would have produced here the same disastrous consequences that ensued in his own.

"If such has been the result of protection upon the general market of the country, its effects have been still more striking, when examined with respect to

The above facts were furnished by members of the Committee residing at Pittsburgh, who vouch for their accuracy.

One fact there stated suggests the following remarks to the Committee:

To the Report of the Select Committee of the Senate of the United States, on the subject of iron, is appended, among other papers, one in which it is stated, that "it is now ascertained that the superiority of England over France is entirely due to the cheapness of iron: a six horse steam engine, for instance, in France, costs, on the average, at least 500 *dollars more than in England*, owing to the cheapness of iron in Great Britain. *It is still dearer in the United States than in France.*"

Here it is asserted that a six horse power steam engine costs 500 dollars more in France than it does in England, and that it is still dearer in the United States than in France. Now it so happens, that in the United States, at Pittsburgh, a steam engine of that power can be put up, ready for action, for the identical sum of 500 dollars.

Prices of Iron at Cincinnati.

In 1814 to 1818, bar iron 200 to 220 dollars per ton; now 100, 105, 110. The fall in prices has been nearly as follows:

1826, bar iron, assorted,	....	125 to 135	dollars per ton.	
1827, " "	....	120 to 130	"	"
1828, " "	....	115 to 125	"	"
1829, " "	....	112½ to 122½	"	"
1830, " "	....	100 to 120	"	"
1831, " "	....	100 to 110	"	"

Castings, including hollow ware, 1814 to 1818, 120 to 130 dollars per ton; present price, 60 to 65, and the quality much improved.

particular, but most important districts. Our Western brethren, the hardy pioneers of our country, were restrained and limited in their contest with the wilderness, by the difficulty of obtaining, on almost any terms, this article, so indispensable to their success in every stage of their arduous enterprise. The second statement exhibits the prices of iron of various descriptions, at different periods, at Pittsburgh and Cincinnati—the great marts of the West. Comment can scarcely be necessary upon the facts there disclosed. The decline in price (in some instances more than half) has been in exact proportion with the stability given to the domestic manufacture, by additional impost on the foreign, until it has reached a point that now enables the mechanics of the first-mentioned city to enter into successful competition with those of almost any other quarter, in the fabrication of nearly every article of necessity, and in one justly esteemed the proudest effort of human ingenuity, they have attained a degree of perfection which enables them to challenge a comparison with the skill and experience of any nation whatever—as the Committee are assured that contracts can be made for any number of steam engines at the prices indicated in the table referred to, it cannot be necessary that they should press this point further. Here your Committee might leave this branch of their subject, satisfied with having, as they believe, demonstrated from facts that protection to the manufacturer, when effectual in amount, and connected with such an assurance of permanence as stimulates enterprise and excites skill, does not operate as a tax on the consumer, but the very reverse. But they believe that facts will justify them in going even further, and will enable them to maintain the position that an impost may, under some circumstances, operate as a tax upon the producer of the foreign article, compelling him, for the purpose of preserving even partial possession of the market, to reduce his own profits in proportion to the increase of impost, which reduction is in fact a contribution to the treasury of the importing country, and may relieve its citizens from the burden of taxation to that extent. The circumstances under which this may occur are twofold. First, when the importing country is the only, or the principal market for the article in question, and that article one which the exporting country must produce in the manufacture of some other article of greater value. Secondly, where there is an increasing surplus of production in the exporting country, and an extensive and growing manufacture of the same article in the country where this surplus has heretofore sought a market.

"Without detaining the Convention longer than to make a passing reference as an example of the first case to the additional duty of the tariff of 1828 (since repealed) upon molasses, which duty was exclusively paid by the foreign planter, who thus contributed, during its existence, more than half a million of dollars per annum to the support of the government of the United States, the Committee will proceed, in proof and illustration, to the second case, and again refer to the statement marked A, and to that marked C,* where, among other facts, it will be seen that in

* STATEMENT C.—The duties on iron imported into the United States, were, 1804 to 1812, 15 per cent.; double war duties from 1812 to 1816. In 1816, duties, rolled iron, 30 dollars per ton; hammered, 9 dollars. The law of 1816,

July, 1828, after the intelligence of our additional duty reached England, iron fell at once $4 44 per ton, and that the following year a further reduction of $6 50 was submitted to. Our additional duty gave additional confidence to the American manufacturer; he extended his operations, and increased the supply without advancing the price. The foreign manufacturer could only reach the market by the payment of the additional impost. American competition prevented him from charging this to the consumer, and

fixing the duties at these rates, ruined many of the manufacturers and compelled them to abandon their works. By act of April 20, 1818, the duty on hammered iron was raised to 15 dollars. This, in some measure, revived the manufacture, and many, who had abandoned, resumed their operations. The foreign manufacturer, to keep possession of the market, offered his iron at a less price; so that there was an actual decline here. In 1824, the duty on hammered iron was raised to 18 dollars, and, in 1828, to $22 40. These additions to the duty had no permanent effect in raising the price. The foreign manufacturer could not advance his prices beyond those of 1824, because the American iron maker supplied the market at those rates; and iron, at a duty of $22 40, sells at less than it did at one of 9 dollars. The foreign manufacturer has been compelled to take the additional duties from his profits; and these deductions from his profits have been paid into the treasury of the United States, without adding to the price paid by the American consumer.

The following table shows the operation of the additional duty levied since 1818 on hammered iron alone:

	Tons.		Duties.	
1818, imported of hammered iron, ...	13,931	....	208,950	dollars.
1819, " " "	16,160	....	242,394	"
1820, " " "	19,531	....	272,877	"
1821, " " "	15,374	....	230,413	"
1822, " " "	26,273	....	378,641	"
1823, " " "	29,014	....	435,210	"
1824, " " "	21,298	....	383,364	"
1825, " " "	23,085	....	428,490	"
1826, " " "	23,837	....	427,066	"
1827, " " "	21,718	....	390,924	"
1828, " " "	33,155	....	663,100	"
1829, " " "	29,202	....	654,141	"
1830, " " " estimated,	29,202	....	654,141	"
Tons, ...	301,880	=	5,369,711	"
Duties at $9, the rate per law of 1816, ...			2,716,920	"
Gain in the treasury, at the expense of the foreign manufacturer, ...			2,652,791	"

he was therefore compelled to diminish his profits to that amount, and to the same extent to become a contributor to the treasury of the United States.

"An examination of the last-mentioned statement C will show that by this means, through the instrumentality of American manufacturers, their foreign competitors have been made tributary to the public treasury upon the article of iron, since that article has been really protected, more than two millions and a half of dollars; while the consumer, as has been already shown, has been benefited to an amount even greater than this.

"If it be alleged that the same benefits would have resulted to the consumer—that the same decline in prices would have occurred, without this competition—we answer that such allegations would be contrary to all experience, which has taught all who have given attention to the subject to know, that while we are dependent exclusively upon a foreign supply for any article of consumption, the foreigner is enabled to prescribe his own terms, and that these always, of course, yield him a large profit. But when domestic skill goes to work to produce the same articles, their prices are reduced, and often to an extent that excites astonishment, when it is accompanied with the knowledge that no sensible change in the cost of production has taken place.

"But it may be asked, if additional protection, by exciting domestic competition, invariably brings down prices, of what benefit is this protection to the manufacturer? To this we reply, that permanence and stability, and not high prices, are our objects. American manufacturers are not so blind to the constant and inevitable course of events, as not to foresee that as these objects are approached, they must expect a more active competition from their fellow-citizens, as well those who are already engaged in the same pursuits, as from others who may be induced to enter upon them. For this they are prepared; they can estimate its extent; and its effects are wholesome and salutary; it stimulates to greater care, economy, industry, and skill: profits are reduced, but they are stable; and the prudent man looks forward with confidence to realizing a fair reward. Against foreign competition there is no guarding, because the manner of its approach can never with certainty be foreseen; nor can its extent be calculated. The ordinary production of foreign industry may, it is true, be calculated with some degree of accuracy, but the extraordinary fluctuations to which their markets are liable from great political convulsions, and other causes, cannot be estimated. Every violent change invariably forces upon our market their vast accumulations, which easily break down the barrier of a mere revenue protection, and involve in ruin all who have essayed competition in the same branch. The consumer receives a momentary benefit, but a reaction certainly follows; great fluctuation engenders a spirit of speculation, and mere gambling is substituted for a regular trade. The frequent occurrence of these evils (everywhere acknowledged to be such) is only to be prevented by an efficient system of protection.

"Having, as they believe, satisfactorily shown the beneficial effects of a system of real protection to the consumer of iron, your Committee will proceed with a few brief remarks upon its influence on the agricultural labor and trade of the country.

"From a critical examination of the returns from 73 furnaces and 132 forges in a great variety of situations, the details of which are more particularly stated in the annexed paper marked D,* they find that in the manufacture of the iron, in its first stages only, made in the United States in the past year, agricultural produce to the amount of nearly three and a half millions has been consumed by the persons thus employed, which vast sum has of course been paid to the farmer, showing how completely his interest is identified with that of the iron manufacturer, and furnishes him with the means of estimating the probable consequences to himself of the destruction of this branch of industry, and the conversion of so large a body of

* STATEMENT D.—The following calculations were made by Hardman Phillips and George Valentine, Esquires, and are derived from the average returns submitted to the Committee, from two counties (those most extensively engaged in the manufacture of iron in Pennsylvania), namely, Centre and Huntingdon, and have been carefully verified by a comparison with returns from 73 furnaces and 132 forges.

For each ton of bar iron and castings made, the following agricultural produce is found to be consumed:

20 bushels wheat and rye,	average at	75 cents,	$15 00
57 lbs. pork,	"	5 "	2 85
43 lbs. beef,	"	4 "	1 72
10 lbs. butter,	"	12½ "	1 25
2 bushels potatoes,	"	30 "	60
½ ton hay,	"	7 "	3 50
For every ten tons of bar iron one horse is employed one whole year, worth $100; and experience shows that the mortality among horses so employed is per annum one in seven, and constitutes a charge, per ton, of			1 43
For fruit and vegetables, of which no return has been made, we feel justified in putting down			1 00
Total,			$27 35

Which, multiplied by the quantity of bar iron and castings, will give the sum of 3,415,850 dollars, paid by the iron manufacturers and those employed by them, to the farmers.

consumers into cultivators and producers. By the same statement it appears that nearly 25,000 workmen are constantly employed, receiving annually the sum of 7,493,700 dollars, making, with their families, nearly 125,000 persons directly dependent on this manufacture.

"For transporting this iron to the markets where it is sold to the consumer, it is estimated that about twelve hundred and fifty thousand dollars are annually paid—being a further contribution to the wealth of the country, through its internal commerce.

"These facts show how completely interwoven are the interests of agriculture, commerce, and labor, with those of manufactures.

"The last consideration that occurs to your Committee, as within their duty to notice, is the capability of the United States to furnish a supply of iron equal to their own wants. Of this the Committee cannot entertain the smallest doubt. The tabular statements, heretofore referred to, show that in two years, from 1828 to 1830, the supply has increased very nearly twenty-five per cent.; and it is known that old establishments, in many situations, are enlarging, and new ones erecting—giving the assurance that this increase will be progressive, until not only the domestic market will be fully supplied, but a surplus remain for exportation.

"If we compare our situation with that of Great Britain less than a century ago, in this particular, we shall see abundant reason for self-gratulation. Ninety years since, her entire production of iron did not much exceed that which is now made in the State of New Jersey. In 1802, within the limits of a single generation, her furnaces were less in number than those now existing in the United States, and the production not more than will be made here during the present year, 1831—and this without availing ourselves of the means to which she is indebted for the extraordinary change which this short period has effected. We have the benefit of her experience; we can command her skill, if it be necessary; we have the mineral and fuels in unlimited abundance (which have done so much for her), when our forests shall fail. Our citizens yield to none in enterprise and ingenuity, when adequate rewards for the exercise of those qualities are held out; and knowing this, with the experience of our rapid progress in the last two years, furnishing, as we now do, more than three-fourths of the entire consumption—is it, we repeat, extravagant to assert that we are fully competent to supply our own wants, and furnish a surplus to minister to those of our neighbors?

"In conclusion, your Committee cannot refrain from the expression of the gratification which the result of this investigation has afforded them. Deserted by the Government, and denied that protection which, at the close of the late war, was freely granted to every other interest, this important branch of domestic industry, so essential to the prosperity, if not to the existence, and so closely connected with our national independence, seemed threatened with absolute extinction. A wiser policy, adopted at a later period, aided by the unconquerable spirit of American enterprise, has raised it to the elevated rank it now holds; and to maintain it in which, it asks no sacrifice from our fellow-citizens engaged in other pursuits. Grateful for the consideration which its well-founded claims for their justice, after years of delay and suffering, at last obtained, it is now returning to the country the full measure of benefit it has received, and will protect them from speculation and monopoly from abroad, should it not again become the victim of that unnatural policy, which would cherish foreign industry while it neglects and destroys our own.

SUPPLEMENTAL REPORT.

"Before separating, the Committee instructed the Secretary to make a further report of any facts that might be received in time for the permanent committee. In conformity therewith, he has now to state that Mr. Peter Townsend, who was delegated by those engaged in the manufacture of iron in this city, to visit all the establishments in this State and those east of it, has returned and reported the result of his examinations; by which it appears that in New York there are in operation, of blast furnaces, 8; in Connecticut, 3; and information from other places enumerates furnaces, not before known, to the number of 26; in the whole, 37 additional furnaces, making, of pig iron and castings, 25,250 tons, and a large number of forges employed in converting the pig into bar iron.

"There are returns of thirty-two bloomery fires in situations were it was not before known that any existed, making 30 tons each per annum, or nearly 1,000 tons yearly of bar iron by this process; and the returns brought by Mr. Townsend show, that the Committee estimated this kind of iron nearly 1,000 tons too low in the districts which he visited. The result of the whole, if 20,000 tons of the above pig iron were converted into bars, would be:

Bar iron	14,285	tons.
Bloomed bar iron, as above	1,960	"
Bar iron, per former statements	96,621	"
Total bar iron made in the United States, agreeably to the information received by the Committee to this date	112,866	tons.

Stated in pig iron it would be:

Former statement	163,542	tons.
Pig iron and castings, as above	25,250	"
Bloomed bar iron, equal to	2,744	"
Total of iron equal to pig iron	191,536	"

"The value of which, according to the mode of estimation already explained, would be 13,369,760 dollars.

"It is hardly necessary to add that these additional facts strengthen all the inferences and calculations heretofore made by the Committee.

REPORT ON STEEL.

"On the subject of steel, your Committee reports, that, as no preparation whatever had been made for collecting information antecedent to this Convention in New York, they are not able to supply it from any other source than what is attainable in this city. They have no time left for collecting and collating tabular statements of the quantity of steel imported and manufactured in this country; nevertheless, enough information has been acquired as to satisfy the Committee that the article in question is one which requires the continued *protection* of Government.

"Without seeking further, the members of your Committee are enabled, from their recollection, to enumerate fourteen steel furnaces in the following places, viz.: Pittsburg, 2; Baltimore, 1; Philadelphia, 3; New York, 3; York, 1; Troy, 1; New Jersey, 2; Boston, 1.

"These furnaces are now in operation, and of a capacity sufficient to supply more than 1,600 tons of steel annually—an amount equal to the whole importation of steel of every kind. But it should be observed, that steel, for common agricultural purposes, is not the best, although it is most used, and that American is quite *equal* to English steel used for such purposes in England. American competition has *excluded* the British common blister steel altogether. The price of blister steel is less than it was before 1828, and probably as low as it ever will be—certainly as low as it ought to be, having a just consideration for the manufacturer and his customer. The only steel now imported from Great Britain is of a different and better quality than that just mentioned. It has been the laudable pride of American legislation to advance with the increasing enterprise of the people, and to encourage discoveries of those mineral treasures towards which that enterprise might be profitably directed. The Committee having shown the result of such countenance from Government, in the instance of common blister steel, may be allowed to anticipate the effects of its continuance, and that protection will be hereafter acknowledged as the parent of perfection.

"Steel imported here from all parts of the world, except England (although the German steel is freely employed in some branches of manufactures), amounts to so considerable a quantity that the competition for ascendency, in our market, must rest between that nation and this. We already supply ourselves, to her exclusion, with common steel; and, to give some idea how extensively it affects our manufactories, the Committee will state two or three striking facts. The iron of this country, when properly made, has been found equal *in quality* to the Russian and Swedish iron used in England for conversion into steel; and, being so converted, is employed in making large and rough implements of manufacture and agriculture.

"It is used in the fabrication of ploughshares; it is worked up by shovel makers, scythe makers, and cross-cut and mill saw makers use more than any other manufacturers. One factory of this kind, in Philadelphia, requires a ton and a half of steel per diem, for every working day of the year. These isolated instances may give some idea of the vast consumption of steel in the numerous factories of the United States, and for this purpose alone they are stated.

"The English, however, continue to supply us with the superior qualities. These are—

"1. Blister steel, from iron of the Dannamora mines of Sweden.

"2. Shear steel of the same origin.

"3. Cast steel.

"As to the first, being the best quality of blister steel, a house in Hull monopolizes all the iron made from Dannamora ore, under contract, by which the parties in Sweden are to forfeit £10,000 sterling if they sell to anybody else ; so that no other European country can furnish a good file without resorting to England for the steel that is made of Dannamora iron, this excelling all others in Europe for files and many other instruments. The British manufacturers, aware of the advantages of their monopoly, continue to exact the same price for their steel delivered in America that they did before the duty on the Swedish iron was reduced in England from $28 88 to $6 66 per ton, thus proving that an article whose low duty approaches nearest to no duty (almost free trade), is charged to this country at a rate no less than before the reduction of duty took place in England.

"It is, however, a cause for congratulation here, that iron of similar or equal quality to that which has thrown all the advantages of manufacturing the best articles of cutlery into British hands has been made recently, by improved processes, from the ore of Juniata, and both sides of the line between New York and Connecticut—the latter denominated the Ancrum, the Livingstone, and the Salisbury ore. Steel is now made at Pittsburg, and may be made in New York and Connecticut, bearing a fair comparison with the best hoop L (L) or Dannamora steel that comes from England. No difference is observed where trials have been made, without disclosing to the judges the origin of either.

"The second kind of first quality British steel is called 'shear steel.' This is nothing more than blister steel, drawn, under a tilt hammer, into bars of various sizes used in the fabrication of some articles of cutlery, and the finer kinds of edge tools. England has hitherto monopolized this branch also, from being in possession of the only European steel that would bear the expense of preparation, and from the perfection of her machinery. She has now the honor of transferring a portion of her experience and skill to the United States. Her workmen in steel, wanting employment or adequate recompense for labor at home, continually seek these among us; and it is believed that these may be afforded to such an extent as to yield them support commensurate with their industry, and that ingenious men, who, under other circumstances, might have been compelled to pursuits not congenial with their education, or to be dependants upon public bounty, will become useful citizens instead of idlers and beggars in the land.

"The third kind of steel (best quality) is called 'cast steel,' and this is made from the best blister steel only. There is none made in the United States. Several attempts to make it, with profit, have proved unfortunate.

"The causes of failure were:

"1. The want of best quality blister steel (of which only it can be made) at a reasonable price.

"2. The want, or expense, of crucibles of proper quality, wherein the blister steel is to be melted and smelted.

"The first difficulty may be surmounted by the discovery that iron well made, from the ores of Juniata, New York, and Connecticut, may be converted to the best blister steel; and the second difficulty is believed to be at an end, since the explorations of the present year have disclosed the existence of clay analogous to that of Stourbridge, which is considered the best in the world for crucibles. Centre, Clearfield, and Lycoming counties (Pennsylvania), have yielded large specimens of clay that satisfy geologists, mineralogists, and chemists of the identity of its properties with those of Stourbridge. Clay in the vicinity of Baltimore has been successfully employed in the manufacture of fire-brick, and may probably be used for the manufacture of crucibles for cast steel, if properly prepared. The great impediment to the making of cast steel has not arisen from any mystery in the art, but the want of strength in the crucibles. Black lead and a variety of clays have been tried, but the weakness of these materials has hitherto caused a loss to the manufacturer, because the crucibles made of them would not bear moving when the melted metal was in them (28th). The Stourbridge was the only kind of clay that possessed the requisite qualities of preserving its shape and soundness when exposed to the greatest heat, and its strength and tenacity when moved for the purpose of discharging the melted metal.

"Capital, enterprise, and perseverance will be engaged to bring this desirable material, so indispensable to the finer arts of cutlery and machinery, into market, if protection be continued to the efforts which our citizens are willing to make. If these views are correct, we have steel for agricultural purposes in the greatest abundance; we have steel (shear steel) for nicer purposes, and we have cast steel for the most refined articles of manufacture among ourselves. But this is not all—we may *export* our steel to Russia, Prussia, and France, in competition with England herself, and thus justify the further importation of foreign commodities which we can have the means of paying for. The subject of steel becomes more interesting as our investigation of it advances; but it is believed that the facts and inferences now set forth will suffice to continue the protection already granted, and to procure time for more extensive practical development, which, if realized, will add to the means of domestic employment and beneficial intercourse with foreign nations.

"B. B. HOWELL, *Secretary*.

"NEW YORK, *November*, 1831."

"The following statement shows the annual consumption of iron in this country, as per report made to the Convention of Manufacturers of Iron, lately held in Philadelphia:

In 1830, there were made 202 furnaces	155,348	tons.
Of which were made into castings	28,273	"
	127,075	
Which rendered into bars at 28 cwt. per ton, would yield	90,768	tons.
There were made of bloomed iron	5,853	"
Total bar iron made annually in the United States,.	96,621	"
Annual consumption of bar iron in the United States	130,007	"
Besides what is made from the blast furnaces into castings	28,273	"
Iron consumed in the United States annually	158,280	"

As early as 1830, the railway enterprises of Great Britain, so intimately connected with her iron trade, extended their influence to the United States, and led to the construction of important lines, connecting the

Atlantic seaboard with the great lakes and fertile valleys of the West, creating a demand for iron rails on a scale of magnitude not anticipated by the iron masters of this country, and led, some years afterwards, to the erection of mills, which now produce one hundred and eighty thousand tons of rails per annum, and, with adequate protection, would soon render this country entirely independent of foreign supplies. It therefore presents a subject of deep consideration to the American statesman, whether the United States, with all her resources, shall continue to be dependent upon a foreign country for so indispensable an article as railway iron, when it can be manufactured at home with better materials, and made to contribute so much substantial wealth to the country.

By an act of Congress, passed on the 14th July, 1832, railway iron was admitted free of duty, provided it was laid down within three years from the date of importation ; but after the 3d of March, 1843, it was made to pay the same duty as rolled iron.

STATEMENT OF DUTIES REMITTED ON RAILWAY IRON IMPORTED INTO THE UNITED STATES FROM 1833 TO 1841.

Years.	Value.	Years.	Value.
1833	$202,210	1838	$910,111
1834	421,010	1839	672,376
1835	529,509	1840	688,510
1836	234,194	1841	391,264
1837	407,517		

From 1832 to 1839, there was no extension of the iron trade ; but the introduction of the hot-blast and other improvements produced the most wonderful results, not only in the increase, but in the economy of the production of iron, by which many large establishments were rescued from ruin. The puddling process was generally adopted, and worked with great practical skill; and this accurate knowledge soon enabled the Eastern establishments to work the American pig metal into nail plates, that have almost excluded and shut out the Swedish iron. The rolling mills were also greatly improved in their machinery and adapted to the higher descriptions, or qualities, and smaller sizes of iron. The blast-furnace, independent of the hot-blast, was much improved and its economy greatly extended by the introduction of cast iron blowing cylinders, water tuyères, kilns for charring wood, etc. During this period of great depression in the iron trade, the United States consumed more than one-fourth of the whole exports of Great Britain.

IRON IMPORTED FROM RUSSIA FROM 1821 TO 1851.

Years.	Hammered Iron in Bolts and Bars.	Value.	Sheets.
1821	84,461 cwt.	$274,593	
1822	120,890 "	383,915	
1823	114,013 "	354,613	
1824	111,957 "	317,843	
1825	79,309 "	241,442	
1826	120,848 "	414,948	
1827	86,325 "	268,216	
1828	117,581 "	366,059	
1829	23,663,505 lbs.	709,827	
1830	21,912,702 "	541,445	
1831	12,720,534 "	314,484	
1832	29,252,017 "	660,459	64,234 cwt.
1833	21,943,500 "	573,280	58,170 "
1834	6,905,680 cwt.		41,760 "
1837	5,240,000 "		40,600 "
1838	5,400,000 "		36,590 "
1850	757,956 "		
1851	793,054 "		

IRON IMPORTED FROM SWEDEN FROM 1821 TO 1829.					
Year.	Cwt.	Value.	Year.	Cwt.	Value.
1821	220,260	$710,392	1826	319,385	$1,078,705
1822	339,885	1,651,055	1827	327,540	979,921
1823	419,958	1,233,826	1828	480,046	1,556,495
1824	333,723	909,857	1829	327,760	1,008,363
1825	399,775	1,274,645			

Although anthracite coal was discovered in Pennsylvania, soon after the Revolution, yet no attempt was made to use it in the smelting of ores, until about 1820. The first successful attempt to generate steam with it was made in 1825, at the Phœnixville Iron Works. Experiments for using anthracite in blast furnaces were made by the Lehigh Coal Company in 1830, and vast sums had been expended, from time to time, in different parts of Europe, to effect the same object; but every attempt proved unsuccessful, until Dr. Geisenheimer, of New York, in 1833, obtained a patent* for smelting iron ores with anthracite coal

* In this patent, Dr. G. claims, *first*, the application of anthracite coal, exclusively or in part, in deoxidating and carbonating iron ore.

Secondly, The application of anthracite coal, exclusively or in part, in combining iron, in a metallic state, with a greater quantity of carbon: if bar iron, for steel; if pig or cast iron, for a superior quality.

Thirdly, The smelting or reducing of iron ore, so deoxidated and carbonated by the application of anthracite coal, as aforesaid, into pig or cast iron.

Fourthly, The refining or converting of iron ore, so deoxidated and carbonated by the application of anthracite coal, as aforesaid, into malleable or bar iron.

Fifthly, The application of anthracite coal in smelting or reducing iron ore, raw or roasted, but not prepared by a previous separate process of deoxidation and carbonation, as above described, into pig or cast iron.

Sixthly, Though I cannot, and do not, claim an exclusive right of the use of heated air for any kind of fuel, nevertheless I believe to have a right to claim, and do claim, the use of heated air applied upon, and in connection with, the said principle and manner discovered by me to smelt iron ore in blast-furnaces, with anthracite coal, by applying a blast of air in such quantity, velocity, and

and hot-blast which he afterwards sold to Mr. Crane, of Wales, who with some additions, took out a patent in this country, in November, 1838, but which he never enforced, for the great secret seems to have been imparted simultaneously in England and the United States.

In 1836, the legislature of Pennsylvania passed an act for the encouragement of the manufacture of iron by mineral fuel, giving to the Governor authority to charter companies with ample powers, in regard to the amount of stock, and quantity of land, for the purpose of prosecuting this branch of industry.

Coke was employed, a few years before the Revolution, in the manufacture of pig and refined bar iron; but it soon fell into disuse, on account of the facility of obtaining charcoal, which made better iron. In 1836 and 1837, the use of it was again revived at Karthaus and Farrandsville, on the west branch of the Susquehanna River, where pig iron was produced, but not of good quality. At Lonakoning, on St. George's Creek, Maryland, the manufacture of pig iron with coke was attended with more success; and at Jennings' Run, in the same coal basin with Lonakoning, two large furnaces, on the Welsh plan, were also erected for using coke or bituminous coal in the smelting of ores.

Messrs. Baughman, Guiteau & Co., put in blast an anthracite furnace at Mauch Chunk, in August, 1838, which worked well for five or six weeks; but for want

density, or under such pressure, as the compactness or density and the continuity of the anthracite coal requires, as above amply and fully described and illustrated.

NEW YORK, *November* 21, 1833.

of ore, it was blown out in January, 1839. The average temperature of the blast was 400° Fahrenheit.

It was again put in blast in July, 1839; and, for about three months, it produced good Nos. 1, 2, and 3 quality, pig iron. The Pottsville furnace was also put in blast in the same year, charged with nothing but pure anthracite and iron ore, and proved entirely successful.

The first account of Mr. Crane's experiments in Wales with anthracite reached this country in 1837; in which he remarked, "Since I have adopted the use of anthracite coal, combined with hot air, the produce of my furnace, with a pressure of 1¼ lbs. per square inch, has ranged from thirty to thirty-six tons, weekly. With respect to the quality of the iron produced by the combination of hot-blast and anthracite, the result is very satisfactory. It is well known that my cold-blast iron, for all purposes where great strength was required, was never deemed inferior to any smelted in South Wales; but that which I have produced with hot-blast and anthracite is decidedly stronger." The entire success of Mr. Crane, at the Yniscedwyn Works in Wales, in 1839, with anthracite coal and hot-blast, gave birth to numerous establishments both in the anthracite and bituminous regions of this country and Wales. In France, experiments had been made under the most favorable circumstances that money and science could furnish, and failed.

TABLE OF BLAST FURNACES APPLIED TO THE MANUFACTURE OF IRON WITH THE ANTHRACITE OF PENNSYLVANIA.

Name and Situation of Furnace.	Name of Proprietors.	When built.	Date of commencing blast with anthracite.	Nature of Ores used, Locality, and Character.	Yield per cent.
1. Mauch Chunk	Baughman & Co.	1838	August 27, 1838	Hematite and magnetic, of N. Y.	40 to 70
2. Pottsville	Marshall, Kellog & Co.	1838	July 10, 1839	Carbonate and hematite	25 to 50
3. Roaring Creek	B. Patterson & Co.	1838–9	May 18, 1840	Fossiliferous, Bloomsburg	50 to 64
4. Phenixville	Messrs. Reeves	1837	June 17, 1840	Hydrated peroxide	38 to 50
5. Danville	Biddle, Chambers & Co.	1838–9	April, 1840	Calcareous peroxide, of Danville	45 to 60
6. Crane Works	Crane Iron Co.	1839–40	July 4, 1840	Hydrated peroxide, near Works	40 to 55
7. Columbia at Danville.	Geo. Patterson	1839	July 2, 1840	Calc. fossiliferous peroxide	45 to 60
8. Montour at Danville	Biddle, Chambers & Co.	1840	July 11, 1841	Fossil calcar's and silicious peroxide	33 to 60
9. Stanhope, N. J.	Stanhope Iron Co.	1840–1	April 5, 1841	Magnetic, of Irondale	50 to 70

No. 1. The Mauch Chunk was the first successful furnace in smelting iron with anthracite in the United States.
2. The Pottsville furnace is the same with which Mr. Lyman made his experiments with a continuous blast of three months.
3. The Roaring Creek furnace, on the north branch of the Susquehanna, was selected for the purpose of experimenting with the rich fossiliferous ores of Bloomsburg and anthracite of Wilkesbarre.
4. The Phenixville furnace experimented with ore from the Yellow Springs, and anthracite from Pottsville.
5. The Danville furnace used the anthracite from Wilkesbarre, and the fossiliferous ores from Montour Ridge.
6. The Crane Works used the ores near their works, and anthracite from Mauch Chunk.
7. The Columbia furnace used the ores from Danville, and anthracite from Shamokin.
8. The Montour furnace used the ores from Montour Ridge, and anthracite from Wilkesbarre.
9. The Stanhope furnace used the magnetic ores of New Jersey, and anthracite from Beaver Meadow.

STATEMENT SHOWING THE POPULATION, AND VALUE OF PIG, CASTINGS, AND WROUGHT IRON IN THE UNITED STATES, FOR 1840.

STATES.	Population in 1840.	Manufactures of Pig Iron.	Manufactures of Iron, castings.	Manufactures of Iron, wrought.
Alabama.......	590,756	$750	$27,700	$4,875
Arkansas.......	97,574	—	1,240	—
Columbia, Dist. of	43,712	—	68,000	—
Connecticut	309,978	162,375	1,733,044	235,495
Delaware	78,085	425	10,700	29,185
Georgia........	691,392	12,350	5,350	—
Illinois	476,183	3,950	41,200	—
Indiana........	685,866	20,250	14,580	1,300
Kentucky	779,828	730,150	164,080	236,405
Louisiana	352,411	35,000	—	88,790
Maine	501,793	153,050	56,512	—
Maryland	470,019	221,900	312,900	513,500
Massachusetts....	737,699	233,300	1,798,758	390,260
Michigan.......	212,267	15,025	57,900	—
Mississippi	375,651	—	36,900	—
Missouri	383,702	4,500	60,300	7,670
New Hampshire.	284,574	33,000	136,334	8,125
New Jersey, ...	373,306	277,850	405,955	466,115
New York,.....	2,428,921	727,200	2,512,792	3,490,045
North Carolina..	753,419	24,200	16,050	62,595
Ohio	1,519,467	880,900	784,401	485,290
Pennsylvania ...	1,724,033	2,459,875	1,262,670	5,670,860
Rhode Island ...	108,830	103,150	147,550	—
South Carolina..	594,398	31,250	—	75,725
Tennessee	829,210	403,213	100,870	628,745
Vermont	291,948	168,575	24,900	42,575
Virginia	1,239,797	470,262	128,256	382,590
Territories. Florida.....	54,477	—	—	—
Territories. Iowa.......	43,112	—	4,000	—
Territories. Wisconsin ..	30,945	75	3,500	—
Territories. Naval service	6,100	—	—	—
Total	17,069,453	7,172,575	9,916,442	12,820,145

No two minerals can at all compare with iron and coal in their unlimited use in all the industrial arts; and until the discovery that anthracite could be used in the manufacture of iron, its immense importance was not fully realized; nor did its great value stop here, for it afterwards led to the construction of some of the most important internal improvements of Pennsylvania, New York, and New Jersey, which have cost, in the aggregate, upwards of eighty-five millions of dollars. The total number of tons of anthracite coal which passed over them in the year 1856, amounted to 6,751,542, of which the Philadelphia and Reading Railroad took to the Atlantic seaboard nearly one-half.

It is now fifteen years since this great work was put into operation to develop the resources of the Schuylkill region, during which time its tonnage has amounted, although limited in the beginning, to 27,034,596 tons, of which 20,718,056 were anthracite coal, and has earned and paid into the treasury of the company $33,654,435. Its average receipts, during the last four years of panics and revulsions, have been $3,770,474 per annum, which, after the payment of interest, repairs, and sundry charges, left a dividend fund equal to fourteen and a half per cent. on the entire stock of the company.

In 1838–9, the iron trade partially recovered from its depressed condition in consequence of the impulse given to railway enterprises, and the price of iron advanced for several years. But it afterwards fell under the influence of the increased make of Great Britain, which was partly forced upon the American market at less than cost of importation, and embarrassed for a while all those who had engaged in its manufacture, under the belief that Congress would soon abandon the free-trade system, and, with increased duties, sustain and foster this important branch of American industry. Bar iron in England gradually fell from £9 per ton, in 1840, to £4 10*s*. in 1843.

By the census returns, the production of the United States, in 1840, amounted to 286,903 tons; but the report of the home league of New York estimated it at 347,700 tons. The mean of both is probably more correct, say 315,000 tons, which shows an increase of fifty per cent. in eight years, under the most discouraging circumstances of a low tariff and a severe financial revulsion, which swept over the country, and paralyzed both its capital and industry.

The cost of making pig and bar iron, at this period, was, in some localities, as follows—based upon a furnace making seventy-five tons per week (average)—

MATERIALS.	*Cost.*
3 tons anthracite coal (steam power, etc., included), at $2 00 per ton,	$6 00
2½ tons of ore, at $1 50 per ton,	3 75
1 ton of limestone,	75
Labor,	1 75
Incidentals, without charge for capital,	2 00
Per ton,	$14 25
REFINING.	
1½ cwt., waste 7½ per cent.,	$1 05
6½ cwt. coal, at $2 per ton,	65
Labor,	1 05
Repairs, ore, and blast,	50
	$3 25
PUDDLING, ROLLING, ETC.	
21 cwt. of refined pig, at $17 25,	$18 11
18 cwt. coal, at $2 00,	1 80
Stock, 25 cts.; Puddling, $5 50,	5 75
Shingling, $1 00; Rolling, 90 cts.,	1 90
Weighing, 12 cts.; incidental, $4 00,	4 12
Add, rolling, piling, and reheating,	1 50
	$33 18

If steam power, 10 cwt. coal for engine.

The total consumption of pig iron, in 1840, amounted to 411,903 tons.

STATEMENT OF IRON MANUFACTURED IN THE UNITED STATES IN 1840.

STATES AND TERRITORIES.	CAST IRON.		BAR IRON.		Tons of Fuel Consumed.	Men Employed in Mining Operations.	Capital Invested.
	Number of Furnaces.	Tons Produced.	Bloomeries, Forges, and Rolling-mills.	Tons Produced.			
Maine	16	6,122	1	0	285	48	$185,950
New Hampshire	15	1,320	2	125	2,104	121	98,200
Massachusetts	48	9,332	67	6,004	199,252	1,097	1,232,875
Rhode Island	5	4,126	0	0	227	29	22,250
Connecticut,	28	6,495	44	3,623	16,933	895	577,300
Vermont	26	6,743	14	655	388,407	788	664,150
New York	186	29,088	120	53,693	123,677	3,456	2,103,418
Pennsylvania	213	98,395	169	87,244	355,903	11,522	7,781,471
New Jersey	26	11,114	80	7,171	27,425	2,056	1,721,820
Delaware	2	17	5	449	971	28	36,200
Maryland	12	8,876	17	7,900	24,422	1,782	795,650
Virginia	42	13,810	52	5,886	36,588	1,742	1,246,650
North Carolina	8	968	43	963	11,598	468	94,961
South Carolina	4	1,250	9	1,165	6,334	248	113,300
Georgia	14	494	29	0	630	41	24,000
Alabama	1	33	5	75	157	1,307	9,500
Louisiana	6	1,400	2	1,366	4,152	145	357,000
Tennessee	34	16,129	99	9,673	187,453	2,266	1,514,736
Kentucky	17	29,206	13	3,637	35,501	1,108	449,000
Ohio	72	35,236	19	7,466	104,312	2,268	1,161,900
Indiana	7	810	1	20	787	103	57,700
Illinois	4	158	0	0	240	74	40,300
Missouri	2	180	4	118	300	80	79,000
Michigan	15	601	0	0	451	99	60,800
Wisconsin	1	3	0	0	1	3	4,000
Total	804	286,903	795	197,233	1,528,110	30,497	$20,432,131

The compromise tariff was regarded, at the time, as an abandonment of the protective policy, and in a few short years, it produced so great a change in the industrial condition of the country that the production of pig iron fell off to 230,000 tons. In 1840 and 1841, nearly all the manufacturing establishments of the United States were closed, and more than three hundred thousand persons were thrown out of employment. Alarmed at this state of things, Congress now opened their eyes to the depressed condition of the country, and in the following year passed a new tariff, which fixed the duty on

Hammered iron,		$17	per ton.
Rolled,	"	25	"
Pig,	"	9	"

which soon had the effect to restore confidence, give employment to the manufacturing classes, rebuild iron works and furnaces which had been rendered inoperative or unproductive, and infuse life and animation into every department of American industry.

In consequence of the railway mania which prevailed in Europe and this country in 1844 and 1845, an increased demand for iron sprung up, which raised the price of it in England and reëstablished the manufacture of it here, on a firmer basis. For several years the production was unexampled in this country, until it was checked by the tariff of 1846. In the meantime, many improvements had been made in its manufacture. The most important of these was the application of the waste gases, which had heretofore escaped from the top of the furnace, to the heating of the steam boilers of the blowing engines, and to the processes of refining and puddling, instead of other fuel. This invention originated in Germany, but the credit of introducing it into this country, with improvements, is due to Mr. Detmold, who took out a patent for it in 1844, and perfected the application of it to all the various processes of manufacturing iron with anthracite. A great economy is also obtainable by Mr. Wall's patent in the conversion of the cast into wrought iron by the application of a current of voltaic electricity to the crude iron in a state of fusion, whether on the hearth of the blast furnace, in the fused pigs in the sand, or on the metal immediately upon its being run from the finery furnace. The Martien process has for its object the purification of iron when in the molten state from the blast or refining furnace, either by air, or steam, or vapor of water applied from below, so that it may rise up amongst and completely penetrate and search every part of the metal previous to congelation, and prior to its being run into a reverberatory furnace for puddling. By this means, the manufacture of wrought iron by puddling, and the manufacture of steel from cast iron, in the ordinary manner, are stated to be greatly improved. Among, however, the most notable improvements in the manufacture of iron, in recent years, is a combination of many kinds of iron to insure good qualities of each; and another is the admixture of small quantities of other metals. Manganese is found to give a closeness of grain to iron. Calamine (carbonate of zinc) gives increased malleability; and, with certain kinds, it produces toughness and strength; while, if antimony be added to the iron of the surface, it imparts a steely hardness, so that qualities can be induced suitable to the different kinds of service which each is to render.

In 1845, the production of the United States was as follows:

540 Blast Furnaces, yielding	486,000 tons of pig.
950 Bloomeries, Forges, Rolling, and Slitting Mills,................	291,600 tons of bar, etc.
Blooms,........................	30,000 "
Castings, machinery, stove, plates, etc.,	121,500 "

which, estimated at the then market value, stands thus:

291,600	tons of	wrought iron, at $80	per ton,	$23,328,000
121,500	"	castings, at 75	"	9,112,500
30,000	"	bloomery iron, at 50	"	1,500,000
				$33,940,500

to which add the quantity imported the year before, say—

46,000	tons	bar iron, rolled, at $60 per ton, ...	$2,760,000
17,500	"	hammered, do., at $80 " ...	1,400,000
16,050	"	pig, converted in castings, at $75,.	1,950,750
6,570	"	scrap iron, at $35,..............	201,950
4,157	"	sheet, hoops, etc., at $130,.......	540,410
2,800	"	steel, at $335, ...	938,000
		Consumption,.....................	$41,731,610

Every ten tons of pig metal required one laborer and four dependents, which gave 243,000 people dependent upon this interest. In France, the estimate is less than eight tons, and in England, more than thirty-five tons. In this country, the agricultural product consumed is very large—say one-fourth of the value of each ton of iron or castings. Consequently, no limit can be placed to the production of this country, except the want of consumption and demand, provided the tariff should be permanently fixed at thirty per cent., *ad valorem.*

By a report made to the British Parliament in 1845, it appears that the value of the British manufactures consumed by various nations was as follows:

Prussia,	7	cents worth to each person.
Russia,	15	" "
Norway,	17	" "
France,	20	" "
United States,	$4 02	" "

The consumption of iron in the United States, from 1821 to 1829, was twenty-five pounds per head; in 1832, forty-seven pounds; and in 1841, it was almost stationary. Under the tariff of 1842, it increased rapidly; and at the end of 1846, it reached a higher point than ever before known, being almost one hundred pounds per head. With the year 1847–8, the domestic production declined in its ratio to population, and the import increased. In 1849, there was a rapid increase in the imports, and diminution of production; yet the total quantity brought into the country was less per head than in 1846–7. The seaports were subsequently filled with every description of foreign iron, and many of the furnaces and iron works were forced to stop work.

Estimate of the Comparative Cost of Manufacturing Pig, Plate, and Bar Iron in Great Britain and the United States, in 1845.

CRAWSHAY—COST.		UNITED STATES—COST.	
PIG METAL.		PIG METAL.	
Ironstone, . . 3 tons, at 6*s.*,	$4 50	Ironstone, 2½ tons, at $1 50,	$3 75
Coke, ... 1¾ " 8*s.*,	3 50	Coal (Anth.) 3 " 2 00,	6 00
Limestone, . . 1 " 3*s.*,	75	Limestone, 1 " 75,	0 75
Wages, ... 6*s.*,	1 50	Wages, ... 1 75,	1 75
General charges, ... 6*s.*,	1 50	General charges, ... 2 00,	2 00
	$11 75		$14 25
REFINING—PLATE.		REFINING—PLATE.	
Waste, 2½ cwt., ...	$1 47	Waste, 1½ cwt., ...	$1 05
Coal, 10 "	1 00	Coal, 6½ "	65
Refining, ...	37	Labor, ...	1 05
		Repairs, ...	50
PUDDLING—BAR IRON.		PUDDLING—BAR IRON.	
Waste, 3½ cwt., ...	58	Waste, 1 cwt., ...	86
Coal, 18 "	1 12	Coal, 18 "	1 80
Do., 12 "	75	Wages and incidental, . . .	13 02
Wages, ...	6 15		$33 18
	$23 19		

From the above estimates, it will be perceived that the wages of labor constitute the principal difference of cost per ton ; since which they have been much enhanced in value in both countries, while capital has become much cheaper.

STATEMENT exhibiting the quantity and value of Pig, Bar, and Railroad Iron; also the value of all manufactures of Iron and Steel IMPORTED *into the United States annually, from the 30th of September*, 1839, *to June* 30, 1855.

YEARS ENDING	PIG IRON.		BAR IRON.		RAILROAD IRON.		MANUFACTURES OF IRON AND STEEL.
	Cwt.	Value.	Cwt.	Value.	Cwt.	Value.	Value.
September 30, 1840	110,314	$114,562	651,117	$1,827,636	581,838	$1,569,944	$3,184,900
Do., 1841	245,353	223,228	1,388,157	2,721,937	465,069	1,064,960	4,255,960
Do., 1842	373,881	295,284	1,122,821	2,001,793	499,400	1,093,070	3,572,081
9 mos. to June 30, 1843	77,461	48,251	247,140	479,911	193,098	358,921	1,012,086
1844	298,880	200,522	682,729	1,201,915	311,544	446,732	3,313,796
1845	550,209	506,291	951,053	1,926,391	436,249	637,514	5,077,788
1846	483,756	489,573	790,802	2,011,770	117,943	281,077	4,981,463
1847	557,114	554,486	841,166	2,303,759	270,733	680,438	5,201,880
1848	1,032,641	815,415	1,445,124	3,435,627	589,789	1,219,185	6,916,620
1849	2,112,649	1,405,613	2,297,841	4,333,592	1,383,265	2,252,246	5,695,948
1850	1,497,487	950,660	2,412,421	4,403,867	2,840,733	3,738,034	7,078,603
1851	1,344,990	787,524	1,717,496	3,322,857	3,772,516	4,901,452	8,182,838
1852	1,837,474	935,957	1,799,261	3,642,332	4,912,510	6,228,794	8,014,618
1853	2,284,549	1,528,031	2,119,064	5,604,414	5,979,904	10,426,037	9,551,884
1854	3,209,673	2,893,483	1,183,420	3,257,899	5,657,339	12,020,309	10,824,645
1855	1,978,495	1,979,463	2,338,216	5,938,732	2,550,327	3,993,900	9,819,461
Total..............	17,994,926	13,728,343	21,987,828	48,414,432	30,562,257	50,912,513	96,684,571

STATEMENT exhibiting the quantity of Pig Iron, Bar Iron, and Nails, with their combined value; and, also, the value of manufactures of Iron not specified, including castings of domestic produce EXPORTED *annually, from* 1840 *to* 1855, *inclusive.*

YEARS.	IRON AND MANUFACTURES OF DOMESTIC PRODUCE.					
	Pig Iron—Tons.	Bar Iron—Tons.	Nails—lbs.	Value.	Manufactures of.	Total.
1840	27	52	2,403,756	$147,397	$957,058	$1,104,455
1841	95	11	2,387,514	138,537	906,727	1,045,264
1842	29	11	2,156,223	120,454	989,068	1,109,522
9 months to 30 June, 1843	13	7	2,629,201	120,923	411,770	532,693
1844	27	6	2,945,634	133,522	582,810	716,332
1845	2	15	1,353,967	77,669	767,348	845,017
1846	198	115	2,439,336	122,225	1,029,557	1,151,782
1847	82	57	3,197,135	168,817	998,667	1,167,484
1848	16	30	3,157,219	154,036	1,105,596	1,259,632
1849	83	251	3,136,958	149,358	946,814	1,096,172
1850	25	338	3,814,488	154,210	1,757,110	1,911,320
1851	351	215	5,300,866	215,652	2,040,046	2,255,698
1852	105	51	3,637,195	118,624	2,185,195	2,303,819
1853	575	140	3,439,183	181,998	2,317,654	2,499,652
1854	2,239	467	4,236,505	302,279	3,908,071	4,210,350
1855	678	157	5,456,493	288,437	3,465,035	3,753,472
Total..................	4,545	1,923	51,691,673	2,594,138	24,368,526	26,962,664

According to the report of the Secretary of the Treasury, in 1846, the production of pig iron in the United States had reached 765,000 tons, and in 1848, 800,000 tons; but in consequence of the railway speculations in England, and the famine of 1848 in Ireland, the prices of iron again receded, and the production fell off three hundred thousand tons—still showing, however, that anthracite iron could hold its own under low duties, commercial disasters, and other depressing circumstances.

In the anthracite districts of Pennsylvania, the production of hot-blast anthracite iron was unexampled, and in 1847 it had reached 389,350 tons, being nearly one-half of the entire production of the United States.

Pennsylvania is, doubtless, the great iron State of the Union, and its soil yields that metallic product in the greatest abundance. The iron ores which are scattered over its surface are various in quality. The magnetic oxide found in the south mountain between the Delaware and Susquehanna, yields from sixty to seventy per cent. metallic iron; whilst the brown and yellowish argillaceous, or hematite, and pipe ores are extensively worked along the borders of most of the limestone valleys, and yield from forty-five to sixty per cent. The fossiliferous ores in Columbia, Union, Juniata, Huntington, Bedford, and other counties, yield from forty to sixty per cent. Iron ores are also most extensively found in the anthracite and bituminous region, and of the same character with the clay ironstone of England and Wales, which yield from thirty to fifty per cent. The species termed bog ore is found in almost every county of the State, and frequently yields from forty to sixty per cent. The great abundance of iron and coal found throughout the State has naturally induced the construction of numerous iron works, furnaces, forges, foundries, and smitheries, where iron is wrought from a crude state into bars and pigs, and moulded into steam engines, cannons, anchors, chains, cables, nails, scythes, cutlery, etc. The rolling mills of this State, in 1847, produced 40,996 tons of railway iron ; but in two years after, the production fell off nearly twenty thousand tons, in consequence of foreign competition, which reduced the price of rails below the cost of manufacture, and compelled the mills to manufacture boiler plate and cut nails, which were less affected by English competition. A comparison of the production of pig iron in this State in 1850 with that of 1847, shows a decrease of nearly one-third since the passage of the tariff of 1846.

STATEMENT OF PIG IRON MADE IN PENNSYLVANIA, FROM 1828 TO 1847.

Years.	Blast Furnaces.	Forges and Rolling Mills.	Pig Iron Made.
1828........	44	78	24,822
1830........	45	84	31,056
1842........	213	169	151,885
1843........	—	—	190,000
1844........	—	—	246,000
1846........	317	—	368,056
1847........	317	—	389,350
1849........	—	—	253,000

Notwithstanding that Pennsylvania is now the great centre of the iron industry of the United States, there are other States west of the Alleghany Mountains which possess great natural advantages, and promise to become her great rivals in the manufacture of iron, all kinds of machinery, and iron wares.*

* The Report of the Commissioners of Statistics of Ohio gives the following account of the Iron production and manufactures of that State:

In 1840	30 furnaces	produced	28,000	tons of pig iron.
1850	35 "	"	52,658	" "
1857	54 "	"	105,500	" "

which shows an increase of one hundred per cent. in seven years.

"The *manufactures* of iron in Ohio have increased even more than its production. The multiplied and increasing uses in all departments of civilized life, create a constant and pressing demand for all its fabrics. Under this demand, and with the vast and various supply of raw material furnished in thirty counties of this State, the manufacture of iron has progressed most rapidly. In the year 1857, the value of iron manufactures, or the products of iron works, exceeded $7,000,000 (seven millions) in the city of Cincinnati alone, where more than fifty of the large machine shops, foundries, and rolling mills are established; and whence are exported to every State in the Valley of the Mississippi, the products of iron work to the value of several millions of dollars. The following brief table exhibits the progress of the iron business of the last thirty years, and is a fair index to the general progress of the State:

	Hands.	Value.
In 1840	1,250	$1,728,549
In 1850	6,075	5,779,495
In 1857	17,000	7,000,000

"Estimating the average number of women and children to able-bodied men, a population of thirty-five thousand were engaged directly, and (taking into view those necessary to supply their wants) probably fifty thousand altogether in the manufacture of iron at Cincinnati. The imports of pig iron at this place (and iron is brought here from several States) is in the aggregate equal to one-third of all made in the State, and is sufficient to employ twenty large furnaces in smelting ore. The exports of iron manufactures from Cincinnati in 1857 exceeded those of 1847, by nearly one hundred per cent.!

"From the facts above presented, it is evident that both the raw material and the manufacture of iron exist in Ohio to a great extent, and are rapidly advancing. The value of iron products may be thus stated:

Value of Pig Iron made at $30 per ton,	$2,165,000
Manufactures of Iron,	20,000,000

"In this business is employed, altogether, more than one hundred thousand persons. The iron business will undoubtedly increase rapidly, and we look forward to the time when not less than two hundred furnaces will be at work, and their product be at least half a million of tons."

STATEMENT OF THE PRICE OF ENGLISH MERCHANT BARS IN LIVERPOOL, FROM 1837 TO 1849.

Year.	Month.	£	s.	d.	Year.	Month.	£	s.	d.
1837	January 14,	10	10	0	1841	September 3,	6	10	0
	February 24,	10	5	0		November 18,	6	7	6
	March 23,	9	15	0	1842	January 1,	6	10	0
	April 15,	9	15	0		April 1,	6	5	0
	May 6,	9	0	0		May 3,	5	15	0
	June 8,	8	10	0		May 19,	5	12	6
	July 15,	7	5	0		June 18,	5	10	0
	August 1,	6	15	0		August 1,	5	7	6
	August 15,	7	15	0		September 1,	5	15	0
	August 23,	8	10	0		September 8,	6	0	0
	September 15,	9	10	0		October 20,	5	12	6
	October 16,	9	10	0		November 3,	5	10	0
	November 23,	9	5	0		December 3,	5	5	0
	December 23,	9	15	0	1843	March 1,	5	2	6
1838	February 7,	9	10	0		April 3,	5	0	0
	March 31,	9	10	0		June 16,	4	10	0
	June 4,	9	5	0		September 4,	4	15	0
	September 6,	9	10	0		October 3,	5	0	0
	December 14,	9	15	0		December 4,	4	15	0
1839	January 10,	9	15	0	1844	March 4,	5	0	0
	January 19,	10	5	0		April 4,	5	5	0
	February 6,	10	5	0		April 18,	5	10	0
	May 17,	10	0	0		May 1,	6	0	0
	June 12,	9	15	0		August 3,	5	5	0
	September 20,	9	10	0		September 3,	5	10	0
	November 15,	9	7	6		December 3,	5	15	0
1840	January 11,	9	0	0		December 20,	6	0	0
	February 18,	8	15	0	1845	January 6,	6	10	0
	March 31,	8	10	0		February 3,	7	10	0
	May 15,	8	5	0		March 3,	9	0	0
	June 30,	8	0	0		April 3,	10	0	0
	July 23,	7	10	0		May 3,	9	5	0
	August 29,	7	15	0		June 3,	8	5	0
	September 3,	8	5	0		July 3,	7	15	0
	September 10,	8	10	0		September 3,	8	5	0
	November 14,	8	5	0		September 26,	8	15	0
	December 3,	8	0	0		October 3,	9	5	0
1841	April 3,	7	15	0		December 3,	9	0	0
	April 19,	7	12	6	1846	January 3,	9	0	0
	May 18,	7	5	0		February 3,	9	5	0
	June 3,	7	0	0		March 17,	9	0	0
	July 1,	6	15	0		June 18,	8	10	0

STATEMENT OF THE PRICE OF ENGLISH MERCHANT BARS IN LIVERPOOL, FROM 1837 TO 1849.

(*Continued.*)

Year.	Month.	£ s. d.	Year.	Month.	£ s. d.
1846	July 18,	8 15 0	1848	June 16,	6 5 0
	August 17,	9 0 0		September 8,	6 2 6
	December 1,	9 5 0		October 13,	6 0 0
1847	January 2,	9 10 0		November 18,	5 10 0
	February 2,	9 10 0	1849	January 12,	5 15 0
	March 2,	9 5 0		February 9,	6 2 6
	June 3,	9 0 0		February 23,	6 10 0
	August 18,	9 2 6		March 23,	6 12 6
	September 18,	9 0 0		April 23,	6 2 6
	December 3,	8 15 0		May 5,	6 5 0
	December 31,	7 15 0		May 25,	5 10 0
1848	January 18,	7 10 0		June 8,	5 5 0
	February 11,	7 12 6		August 3,	5 10 0
	April 7,	7 5 0		October 26,	5 5 0
	May 12,	6 15 0		November 24,	5 10 0

In 1849, a convention of iron masters assembled in Philadelphia to consult on the best mode of relieving the iron industry of the United States from its depressed condition, and drew up the following memorial, which they submitted to Congress :

TO THE SENATE AND HOUSE OF REPRESENTATIVES OF THE UNITED STATES OF AMERICA IN CONGRESS ASSEMBLED :

"Your memorialists, interested in the manufacture of iron in the State of Pennsylvania, ask leave to offer some considerations and statements suggested by the suffering condition of that industry. We are not unaware of the prejudice which exists in the minds of many, against the propriety of the government giving any attention to the grievances of manufacturers; neither are we ignorant of the grounds of this feeling.

"It is a part of our purpose in this memorial, to lessen, if we cannot wholly remove this prejudice. On a subject of such importance, involving so many interests, in a country so extended as ours, it is to be expected that honest differences of opinion will exist, and sectional, if not clashing, claims will arise. The manufacturers of this country, whatever may be their troubles, must yield with all their fellow-citizens to that system of compromise on which all our institutions are adjusted. We cannot ask any legislation for our advantage unless it be, if not equally for the benefit, at least not injurious to the rest of the community. On this ground we are willing to base our present application for relief. We come, without distinction of party, and ask to be heard upon strictly national considerations, that if any enactment is consequent upon our petition, it may be regarded as permanent and not partial legislation. We ask not for relief to-day which may be withdrawn to-morrow; but, for *a settled policy*. We ask to have the wisdom of all interests and all parties applied to the preparation of such a system as will be permitted to stand, subject only to the improvements which experience and time may dictate.

"It cannot be questioned, that a large supply of iron is necessary to the rapid progress of any country in all departments of industry and the arts, in civilization and the material well-being of the people. The production of iron in Great Britain is equal to that of all Europe beside; while her consumption is equal to a million and a third of tons, or about 100 lbs. to each individual of the whole population. Belgium falls little, if any, short of an equal consumption for each inhabitant. Sweden would stand next in order but that she exports so much of her iron as to remain far behind Belgium in proportionate consumption. France consumes about 30 lbs. for each person, and of this, about one-tenth is imported. The rest of Europe does not consume 10 lbs. each person, and the remainder of the Old World does not reach a consumption of 5 lbs. In this respect, the enterprise and industry of the people of the United States have not permitted them to remain behind; so that despite of obstacles the most formidable and the most vacillating legislation, we stand in the front rank of nations as to the consumption of iron. Our consumption is equal to that of Great Britain for each inhabitant; but we import about two-tenths of the quantity consumed. Such is the abundance of raw materials, such the enterprise of our people, such the tendency to employ iron, and so greatly are the facilities for transportation multiplying, that we might with certainty outstrip the world in its production. All that is needed to secure such a result is a steady home market. Pennsylvania now produces as much iron as Great Britain did in 1820; her product has doubled in ten years, under great disadvantages, and in ten years of favoring legislation, it might be doubled again. Pennsylvania now produces as much iron as France; more than Russia and Sweden united; and more than all Germany. Yet how many States of the Union will ere long manufacture as much as Pennsylvania? for there are few in which the raw materials do not abound. Our population is destined to increase in a very rapid ratio; under a wise policy, the production of iron would far more than keep pace, until we should be finally as much distinguished for the consumption of iron as we now are for the production of cotton.

"The policy of purchasing only in the cheapest market sends not only the people of the United States, but of all the continent of Europe, and in fact of all the world, to Great Britain for iron; for there the cost of making is one-half less than here, and in still greater disproportion with most other nations. The difficulty is, that the manufacturers and merchants of that country are not governed by the cost of production in selling their commodities, but by the extent and urgency of the demand. When there is a demand, the prices are at the highest; when there is not, the world is invited to a cheap market.

"If it be objected to such a development of the manufacture of iron, that the cost of production is too great in the United States, and that we ought rather to import that which is purchased cheaper in other countries; the reply may be made that, Great Britain being the only country in which iron is sold at lower rates than here, our demand could only go to that market; that if sound economy requires us to obtain our supply of iron in Great Britain, the same motive would send all other nations to the same market. But our orders alone could not be filled without so raising the price, as to preclude all possibility of our obtaining a full supply. If we should order from Great Britain in one year, additionally, half the quantity of iron we now manufacture, prices would go higher than they have been for a century, in England or America. The British iron market is cheap when you refrain from it, not when you press upon it. The cost of manufacturing iron is far from being the only, or even the chief controlling element of the price. The manufacturers and holders of iron in Great Britain are extremely sensitive to a demand for any increased quantity of iron or to any increased urgency of demand, whether from abroad or for home consumption.

"A million of tons of iron, which is the amount of our consumption when the industry of the country is suffering under no depressing causes, would have cost in Great Britain, in 1843, at the prices then prevailing (taking half the amount as pig and half as bar iron), £3,500,000 sterling. In 1846, the same quantity would have cost £9,000,000 sterling, at which price it was more economical to manufacture than to import. These high prices gave an immense impulse to the production of this country, and showed how promptly capital and enterprise combined to overcome an emergency by which the country was threatened with a deficiency of the indispensable article of iron.

"Had we even a stipulation, by treaty, on the part of the government of Great Britain, that we should always be furnished with iron in that market at the low rates now current, say a million of tons for $20,000,000, how could we pay for it? We already import more than we can pay for in exports.

"All the shrewdness and enterprise of our merchants are constantly at work to increase our exports; not only is everything exported that will pay a profit, but every article that will pay a freight. How absurd to suppose we could pay $20,000,000 additional for iron! Any attempt to supply ourselves with iron from abroad would, if persevered in, reduce our consumption from 100 lbs. for each person to far less than half that quantity, besides abridging our imports of other articles, and wholly deranging our foreign commerce.

"As manufacturers of iron, we freely admit that we enjoy in Pennsylvania, and, we may add, in all the United States, very manifold natural advantages. If we could now boast that exemption from injurious rivalry enjoyed by the British manufacturers, during the rapid growth of their industry, we could safely promise even greater results than have been witnessed elsewhere. Look for a moment at the circumstances under which the British manufacture of iron was developed. There was no surplus of pig iron in any country of Europe, and the article was unknown in European foreign commerce. All that England ever imported was a few thousand tons from the colonies of Pennsylvania, Maryland, and Virginia, and this was finally cut off by our revolution. The English manufacturer of pig iron had no rival, and required no protection. The only competitors in bar iron were Russia and Sweden; their prices, from 1780 to 1849, ranged from £12 to £25 per ton. But as if this high price was not ample protection to British manufacturers, the government advanced the duties fifteen times, between 1780 and 1820, without one reduction, increasing them from £2 10*s.* to £7 per ton, affording the double protection of high prices and constantly increasing duties.

"Between 1780 and 1825, Russian and Swedish bars could not be imported and sold in England for less than £20 or $100 the ton; this gave the English manufacturers entire possession of the home market for all purposes to which their iron was applicable, and yet their price was always below the foreign.

"In contrast with this, the American maker of bar iron competes with rivals whose average home price is only £8 or $40 the ton, and who, at present rates of iron in the British markets, and duties here, can put their bars in our market at $40, duty paid. It is true, they lose money by the operation, but they would lose more by selling at home and thus further depressing the markets in which they must sell three times as much as they export. Thus they preserve their own, and ruin the markets of their competitors. During the rise of this manufacture in Great Britain, pig iron was worth in their market over 100 shillings, generally 120 shillings. The American manufacturer encounters pig iron sold in Scotland, for years together, at from 35 to 45 shillings, and which can now be put down in our markets, duty paid, at 60 to 70 shillings.

"If we ask relief against such ruinous competition, we derive countenance from the fact, that British manufacturers constantly appealed to their government for protection under the favorable circumstances we have noted. We have seen with what success. The time was not long until in 1825, the manufacture having attained ample growth and power, it could dispense with all aid, and defy competition. Great Britain had then risen to the rank of the largest consumer of iron in the world.

"If this business has been overdone in Great Britain, the evil consequences have fallen upon the manufacturers. The public has enjoyed an immense advantage in the abundance of a material so important in every department of industry as iron. The fluctuations in price which have ensued from this large production have been of late years so great as to cast in the shade all other commercial changes of price. The range of these fluctuations in pig iron during the last ten years is from £1 18*s*. to £5 12*s*. 6*d*., and in bar iron £4 10*s*. to £13, or about 200 per cent.

"In one extremity of this fluctuation, British iron becomes too high to import under a revenue duty; in the other, too low to admit of home production. In the one extreme, we cannot afford to use it; in the other, it paralyzes our efforts to manufacture for ourselves.

"The legislation asked by American manufacturers deserves not the odium so frequently heaped upon it. We know that we can furnish to the consumers of this country a million of tons of iron cheaper and better than it can be had abroad. We ask for defence against those commercial fluctuations which occur in Great Britain, from causes wholly originating there, and which, while they thrust down the prices of iron there far below the cost of making, throw large and irregular quantities into our ports, disturbing the regular course of industry here ; breaking down our markets, and carrying ruin at each such invasion, into many establishments. If we ask aid against such irregularities, it is no more than we should be obliged to do, if the manufacture in the United States were as greatly developed as in Great Britain, and enjoying, in all respects, equal advantages. If that were the case, each of the equally powerful competitors would seek to relieve their home markets in seasons of depression, by thrusting the rejected surplus upon his rival; and each would seize the opportunity of high prices in the other to make large exports, until both markets, unable to maintain any high prices to compensate for unfavorable periods, would sink into hopeless depression, and the business perish or be greatly impaired. Against such consequences both would appeal to their respective governments for protection, not for monopoly ; for that security against ruinous fluctuations, and that regularity in sales indispensable to the success of industry. Competitors at home can observe their mutual progress, and take their measures of defence in time, but that competition which comes from abroad, cannot be watched, nor preparation made for its sudden inroads. If the British manufacturer is prevented from flooding our markets at less than the average upon which his business thrives, a mere revenue duty will be ample protection against the great advantage he enjoys, of employing labor at less than half the cost paid in the United States.

"Among those most deeply interested in the vigor

and prosperity of our iron manufactures are the farmers who furnish food, and the planters and manufacturers who furnish clothing, for our operatives in iron. We cannot here fully unfold the chain of mutual interests which binds all branches of industry together, nor exhibit its strength, and the importance of preserving it unbroken. We ask attention to only a few prominent facts. When the ports of Great Britain were opened to our agricultural products, it was fondly hoped that our farmers would find there an unlimited market for wheat and maize. At the present moment, however, these are very little higher in Liverpool than in Philadelphia, and the pressure of any increased export would sink prices there below ours. At the present rates of iron and flour in Liverpool, the flour made from an acre of good wheat will about exchange for a ton of pig iron, and pay for its transportation to this country. If we take the product of the acre at four barrels, worth now in our market $18 or $20, it will exchange here for a ton of pig iron of far superior quality.

"But farmers who feed the manufacturers of iron in the United States do much better than exchanging the product of an acre for a ton of pig iron. A furnace yielding 4,000 tons of pig iron gives employment to two hundred laborers, each of whom consumes annually fifty dollars worth of food. Of this but one-tenth is expended for bread; the remainder is consumed in the shape of mutton, veal, pork, beef, poultry, potatoes, turnips, beets, and other products of garden, field, and orchard; the production of which in great variety is an accompaniment of all good husbandry and profitable farming. To import 100 tons of pig iron requires the product of 100 acres of wheat; but in our home markets the usual product of 50 acres will exchange for 100 tons of pig iron. An acre of potatoes, the cultivation of which does not exceed in expense that of Indian corn, will exchange for eight tons of pig iron in the markets of Philadelphia. The farmer who, with 100 acres of wheat, prefers the foreign market, will receive for his crop 100 tons of pig iron, at present rates worth $2,000, whilst he who has a hundred acres of potatoes can exchange his crop at home for 800 tons of iron, worth $16,000.

"Wheat sent to a distant market, which fluctuates according to the supply and demand, must be sold without reference to the cost of production, and without control of the producer for what it will bring in competition with all the world. What the farmer sells at home is at his own price, and is sold or held according to his discretion. Well-cultivated lands dependent on a foreign market may be worth from $5 to $20 per acre; those that have the full advantage of a home market are worth from $50 to $200. If the production of iron in Pennsylvania were continued in full activity for ten years, it would double the value of her own lands and make a vast contribution to the value of other lands and property beyond her boundaries.

"What is applicable to the propriety of sending wheat to a distant market to be exchanged for iron, is just as true applied to the expediency of sending raw cotton to England, to be exchanged for manufactured cotton, or any other foreign goods. The cotton plantations can feed the operatives necessary to manufacture all their cotton; and such a policy would triple the value of every cotton plantation in the country. To produce this additional quantity of food would probably require no more laborers than are now employed in growing cotton. It would only require that division of labor which is as important to the success of the planter and farmer as to that of any other producer.

"To manufacture 800,000 tons of iron, the present product of the United States, gives support to upwards of 250,000 persons, to whom at least twenty millions in wages must be paid. Of this sum $4,000,000 will be expended in coarse cotton fabrics for clothing and furniture, $3,000,000 for woollens, and $3,000,000 for other items of clothing and domestic comfort. The $20,000,000 earned by the operatives in iron will thus be diffused over the whole country, giving vigor and activity to numberless branches of industry. The South will furnish cotton, sugar, and rice ; the middle States, bread, potatoes, and meat ; and the northern States, the products of the loom ; whilst thousands of tailors, hatters, shoemakers, and other tradesmen, find constant employment in ministering to the necessities of the makers of iron, consuming themselves an additional quantity of food and clothing by a demand distributed in like manner.

"It is said the domestic cost of manufacturing iron is too high to be sustained by any sound legislation, or to warrant any large consumption. We reply that our whole supply cannot be imported as cheaply as we manufacture it ; for the reason that the cost is not the only controlling element of price, and that our large demands, if made upon the British market, would quickly enhance prices far beyond the domestic rates. We must, therefore, manufacture at home at least three-fourths of our consumption ; and, to do this, our manufactures must be maintained in full vigor by remunerating prices and a steady market. Iron costs twice as much to manufacture here as in Great Britain ; because employers here pay double, and more than double, for wages of labor. The laborers of the United States can be fully employed at the high wages which prevail here, and we are not prepared to say that these wages are more than a just compensation for labor. It is certain, that in most countries where less rates are paid, a large mass of the population is in a state of destitution, and sunk to the lowest grade of human existence. In this country, where physical well-being is so easily attainable, should we not feed, clothe, and lodge our laborers in comfort, and keep them out of the poor-house ? The wages now paid are only sufficient for this, and to enable the prudent to make some savings for sickness, reverses, and old age. We are not, therefore, in favor of any system which contemplates a reduction of wages, and a consequent degradation of our working men. We believe that the consumption of every country is regulated by the wages of the laborer : if he is liberally paid, he will consume freely. The mass of the consumers in a country must be the laborers ; and, when these are able to exact a fair compensation for their toil, all prices must soon be adjusted upon the same scale. The manufacturer will demand for his product a price proportioned to the cost of labor ; the farmer must do the same, and so on through the whole circle of industry. The laborer himself contributes to sustain these prices by a consumption proportioned to his income. All persons concerned in this adjustment being in a condition to ask and obtain justice, the whole system of consumption will be regulated by the rights of all and the means of all. In this state of things, the largest possible consumption can take place; because it will be the result of a fair exchange. The stimulus to exertion and increased production will be complete, because every product of industry can be exchanged, at a fair rate, for other products. If no disturbing cause intervenes, the production and consumption need have no other limit than the physical ability of the producing parties and their mutual wants.

"In full activity of business in the United States, our consumption of iron has reached 100 lbs. for each person. If no disturbing cause had interfered, we should now be consuming 200 lbs. Our farmers could amply feed the laborers needful to such an increased production, and our machinists and mechanics could soon, under the operation of such a system, work up and prepare it for consumption. Every branch of industry would have all the rest for customers; and, if all measured their values by the same scale, all would be rewarded according to their industry. It is well known that low prices of iron are no boon to those who buy to work up and sell, and that the seasons of highest prices are often periods of largest consumption. In 1847, pig iron ranged above 30 dollars per ton in this country, yet at these high prices the whole stock of that year, estimated at 750,000 tons, was consumed; all the old stocks and remnants were swept off, and it was perfectly apparent to those well acquainted with the state of the market, that there was an actual deficiency of supply to the extent of very nearly, if not quite, 100,000 tons. In 1849, with pig iron at 20 dollars, and bar iron at 50 dollars, the consumption of the country has probably fallen off one-third, and the production one-half. With this diminished production, domestic stocks are now accumulating rapidly. Of the amount imported this year, a very large proportion yet remains in the market. The quantity of iron now on hand in this country is estimated at 300,000 tons; and of this, one-half is British. The manufacturers of castings, of machinery and hardware, now find that the consumption of their articles is checked, and that the low price of their raw material is not only no benefit, but a positive evil, and they are ready, equally with the makers of iron, to ask for a remedy. A similar result will be found by comparing all the periods of high and low prices.

"To whom, then, inures the advantages of cheap foreign iron? Abundance of food is no more beneficial to a man in the agonies of a fatal disorder, than cheap iron to a paralyzed industry. The ability of the country to consume iron depends on the vigor and activity of all departments of industry. If agriculture languishes, the consumption of iron is diminished; if the machinery of the North is idle, or partially so, the demand for iron falls off; and so, if cotton or sugar are selling at inadequate rates.

"At the present moment, various interests are suffering from the utter stagnation of the iron trade, as the operatives in iron will this year, 1849, consume in supply of their wants some twelve millions of dollars less than in 1847. This alone is enough to carry serious injury into numberless channels of industry.

It especially affects the consumption of cottons and woollens; for the use of these can be abridged to a greater extent than food. All interests are, therefore, bound together by common ties; when one suffers, all suffer. It is a great mistake to suppose that the producers of cotton, sugar, rice, and tobacco, have no special interest in the activity of manufacturing industry in other States. A very large proportion of the cotton crop is now consumed in the United States, and thus kept from the British market, already so liberally supplied as to give British merchants control of the price. When British iron is exported to us for want of a market at home, we take it at our own price; when we order large quantities of iron, we pay what they can exact. Our cotton is mainly exported, disgorged upon the British market, and the price is made in Liverpool. When British manufacturers shall be compelled to come hither for their cotton, the price will be made by the planters. The present supply is so large, that the price is yearly the result of mere speculation. What is sold in this country is clear gain to the planter, as the whole crop would sell for no more in Great Britain than the quantity which now goes there. If half the crop were consumed at home, the other half would sell for as much in Great Britain as is realized for the quantity now exported. This result is not only attainable, under favoring legislation, but it might have been attained before now, by that wise policy which stimulates home industry to its utmost capacities. By such a policy, the consumption of cotton and iron could be doubled in a few years, with immense advantage to the wealth and happiness of the whole population. It is the interest of the planter not to struggle for that division of labor among nations; which makes one nation a planter of cotton, another of sugar, another a maker of iron, another a spinner, another a weaver, another a tailor, and so on; but that division of labor which mingles these pursuits in the same country, in the same county, in the same town, and, to some extent, on the same plantation. This is the division of labor which begets a vast production and consumption at home, and an internal trade with which no foreign commerce can ever vie.

"Who can doubt, that if the planting States were legislating for themselves, their first care would not be to become more independent, to diversify their labor and vary its products? What such legislation would compel them to do, they can now do under that national legislation which is invoked by others. They are already entering upon that career—it will be found not only the sure road to prosperity for them, but also for us. We so fully confide in the doctrine of the division of labor at home, that we not only trust that the cotton planters will manufacture as much of their cotton at home as they can, and feed the operatives thus employed, but also manufacture as much of their iron as they can. There is room for all, work for all, and market at home for such a large portion of our products that the remainder will not overcharge the channels of foreign commerce, and be sacrificed for the advantage of foreign merchants and manufacturers.

"We object to the doctrine that industrial pursuits are subordinate to foreign commerce; and that the latter is to be considered as the rightful patron of industry. In our view, industry stands first in natural order, and should be the first care of the legislator. Commerce is merely an agency, the charges of which, as well as its powers, should be kept to the lowest point consistent with efficiency. It may suit those engaged in commerce to insist upon the 'Let us alone' policy, for, doubtless, merchants can take care of themselves, and thrive not the less, when the producers, from whom their profits come, are suffering most. The manufacturer has, in all countries, asked for special legislation, and under its good effects, the present manufacturing system of Europe and this country have grown to their present magnitude. The relative importance of the domestic production of this country and its foreign commerce, may be seen in the fact that our foreign commerce yields from six to eight dollars' worth of foreign commodities to the consumption of each individual of our population; whilst the domestic industry of the country furnishes not less than from 75 to 100 dollars for each person. Shall we pursue a policy impairing the power that produces the larger supply, in the vain attempt to add the worth of a dollar or two a head, to the quantity of foreign commodities consumed? And be it noted, that every dollar a head added to our consumption of foreign goods adds over 21,000,000 dollars to our imports.

"If an ample supply of iron be indispensable to national progress and national welfare, and if the whole of that supply cannot be imported as cheaply as it can be made at home, the principle which should govern legislation applied to this industry, and to others in like circumstances, is clearly discernible.

If home production, on which we rely for more than three-fourths of our consumption, is not sustained in that activity which insures its proceeding with economy and advantage, it must flag; and the product being diminished, a greater demand must be thrown upon the foreign market, enhancing the prices of importation. But if the home production is adequately sustained by a free market, it can supply all the channels of consumption. Legislation, marking closely the line of vigorous production at home, will encourage importation, with the double purpose of obtaining revenue, and keeping the manufacturers at home to fair prices.

"Sustain the domestic manufacturer at the point of full production, and then admit the foreign article freely. The more closely our revenue enactments approximate this object, the more perfectly will they encourage domestic industry, obtain the largest attainable revenue, and best secure the interests of consumers. The manufacturer, constantly struggling to keep up his prices, will be as constantly met by foreign iron, selling at such rates as to keep him to the line of public advantage. It is the operation of a well-managed competition between the domestic and foreign producer, which results in the greatest benefit to the consumer. If the consumer is driven to a foreign market for his supplies, or for too large a proportion of them, prices will be inordinately advanced against him; while, if the foreign market is prohibited, or too heavily burdened, the same undue advance may take place at home. But if foreign iron is introduced at the point designated, it not only works no injury, but produces positive public good, as to revenue and prices, and also as to the increased consumption of iron. There are certain average rates at which manufacturers of iron in this country can live and flourish, and these rates are very little, if any, above those to which the often recurring fluctuations of prices in Great Britain are carried. At these rates, which are easily ascertained by the legislator, the line of competition can be established, with the greatest advantage to the consumer. They will not exclude foreign iron, but frequently attract it. During the last fiscal year, the very large importation of 315,000 tons of iron has taken place. Of this, much the larger proportion has probably been sent to us on foreign account, because there was no demand at home; it was sent to save the home market, already broken down, from further depression. It has broken down our markets; and if sold at present rates, will not yield the makers a penny of profit. This iron, coming thus to a bad market, came because it would have been worse for the holders to keep it at home. If previous legislation had shielded our market, so as to maintain prices remunerating to our manufacturers, the additional duty necessary for this purpose would not have deterred the export of iron to this country; for, while those who shipped it to our ports must have paid a higher duty, they would have realized better prices. A ton of iron rails, under the present tariff, at the prices prevailing in 1846 and 1847, was charged with the duty of twenty dollars, which was almost prohibitory, and therefore produced little revenue, making foreign rails cost 90 dollars per ton. During the year 1849, a ton of rails has been charged with only eight dollars, and has, of course, produced but little revenue ; whilst a ton of rails were laid down in our market at 45 dollars, injuring the domestic producer to an extent that is incalculable. A system of revenue which would meet the low prices by a proportionate increase of duty, and make provision for high rates by a like reduction, never excluding the foreign iron, would, we believe, meet the exigencies of domestic industry, and greatly increase the revenue. Whatever may be the advantages of the *ad valorem* system in other cases, they are more than neutralized by the fluctuations of the prices of British iron. It is true that a part of this objection applies with equal force to specific duties ; for, when these are high enough to meet the difficulty of low prices, they become prohibitory when prices rise. These considerations furnish a strong inducement for special provisions in our revenue system in regard to foreign iron. A system could thus be devised which would give a mighty impetus to the production and consumption of iron, and to other dependent branches of industry. A home competition could be thus insured, which would, in the end, reduce the price of iron to the lowest limits consistent with undiminished production. Under such a policy, we should soon surpass Great Britain in the quantity of iron made and consumed, as much as we do now in the quality. We should employ hosts of laborers, and attract them hither from all quarters of the world ; and for every million of people which this scene of industry would draw to our shores, we should be furnished with an additional home market, equivalent in amount, and far more remunerative, than the average export of our foreign trade.

"In closing this memorial, we ask your intervention in our favor, and the insertion of such provisions in our revenue laws as will 'regulate commerce with foreign nations' in iron, and exclude from our markets the results of those destructive fluctuations and irregularities which originate in foreign causes, and should expend their force on foreign shores. This being done, we only ask further that such duties be imposed upon foreign iron as will bring the largest revenue to the public treasury."

RESOLUTIONS.

1. *Resolved,* That a crisis has arisen in the iron business which calls for the immediate revision of the revenue laws, so far as that article is concerned, and that the number of establishments which have already been forced to suspend by the influx of foreign iron, proves that without such a change, the business cannot permanently sustain itself in its rightful position, as a great branch of our national industry.

2. *Resolved,* That the manufacture of iron is not a mere local or individual interest, but is of national importance, as affording a supply of a chief element of progress in time of peace, and an important engine of defence in time of war.

3. *Resolved,* That it has been the policy of every civilized government to extend a fostering care to the production of iron ; and that the example of Great Britain is especially worthy of notice, who by an unwavering course of favoring legislation, sustained the business in its infancy, when its prospects, without that aid, were far less encouraging than they now are in this country, and never removed its protecting care, until it reached a development unparalleled in the history of trade ; by which the consumer is supplied at the lowest rates, and the manufacturer defies the competition of the world.

4. *Resolved,* That in this country the development of its industry and of its various natural resources has been so rapid during the last few years, that the only remaining serious impediments to the establishment of the iron business here on a firm basis, lie in the high price of labor in this country, compared with that which rules in foreign

countries with which we compete, and in the enormous fluctuations which have prevailed in the price of iron.

5. *Resolved,* That experience teaches that the difficulties produced by these causes cannot, in the present state of the trade, be overcome by any unassisted efforts on the part of the manufacturers; that the business is now prostrated, and wide-spread ruin threatened, because they are exposed to them, and that an efficient remedy is only to be found in the action of the general government.

6. *Resolved,* That in raising revenue by duties on imports, the government can readily apply this remedy, by imposing such a rate of duty as shall establish *a fair competition* with the foreign article, and, without *prohibiting* its introduction, build up such a domestic production as shall always keep the price of foreign iron in check; and that it is the manifest interest of the *consumer of iron,* who must in any event be taxed for revenue, to have this tax so framed, if possible, as to prevent the foreign maker from obtaining control of the market and fixing his own price, as he may do in the absence of domestic production.

7. *Resolved,* That the present *ad valorem* duty on iron might, as suggested by the late Secretary of the Treasury, be increased with advantage to the revenue, but that the *ad valorem* system cannot possibly meet the true and mutual interest of consumer and producer, because the fluctuations in price are aggravated, instead of being relieved, by duties which increase as the price rises, and diminish as the price falls; thus subjecting the consumer, on the one hand, to supply his wants at extravagant rates, and the producer on the other, to a hopeless struggle against the cheap labor of Europe, combined with the immense capital accumulated in years of high prices at the expense of the unprotected consumer.

8. *Resolved,* That the *ad valorem* system, moreover, operates precisely as if a new tariff were enacted at every change of price; that it encourages frauds upon the revenue, because a small invoice price produces a duty proportionably low; that it tends to throw the importing business of the country into dishonest hands; that it offers a virtual bonus upon the consumption of inferior articles, and this operates to degrade the national taste, and retard the onward march of improvement, while, at the same time, its practical operation has not developed a single advantage which the "specific system" does not also possess.

9. *Resolved,* That specific duties, on the other hand, tend to coun-

teract the fluctuations of trade, and insure regularity to the business of the country, as the producer can then calculate with comparative certainty upon the competition against which he has to contend, and can establish his production upon so firm a basis, that, on the average of years, as cheap, if not cheaper prices, are insured to the consumer.

10. *Resolved*, That the true test of national prosperity is to be found in the price of labor, which can only be maintained at a permanently high rate by opening every possible avenue of employment to receive the ever-increasing supply ; that the *ad valorem* system of duties tends to narrow the field of labor, and consequently to reduce its reward.

11. *Resolved*, That the employment of labor in manufacturers, by diverting a large amount from agriculture, insures a home market and better prices to the farmer.

12. *Resolved*, That the cotton planter is equally interested with the farmer in securing large wages and diversified employment to labor. because it increases the ability of the community to consume, thus securing at the same time a large home market for cotton, and by reducing the quantity forced upon the foreign market, a higher price for that staple.

13. *Resolved*, Therefore, that the *protection* which this Convention desires, is protection for the *government* against frauds upon the revenue ; for the *importer* against dishonest competition ; for the *producer* against the unnatural fluctuations of trade ; for the *consumer* against the high prices of iron, which are rendered inevitable by the absence of domestic production ; for the *laborer* against the meagre rewards which a narrower field of employment must oocasion ; and for the *farmer* and *planter* against the obvious disadvantages of a single market, and the loss of intrinsic value entailed upon his produce by the expense of transporting it abroad.

14. *Resolved*, Therefore, that it is the sense of this Convention, that stability and permanence in the tariff law, is the first and most essential requisite to a healthy state of the iron trade ; and that, as manufacturers, we prefer the lowest duty that will enable us to prosecute our business, provided it be permanent, to a higher one with a prospect of a change in a few years, and that we are indifferent whether this object is effected by the imposition of specific duties, or the adoption of a sliding scale, so framed as to counteract the fluctnations in price.

15. *Resolved*, That this meeting expressly disclaims all political motives, and all desire to further the views of any party by its action; that it is our opinion that political agitation on the subject of a tariff upon iron should cease; that in all our efforts to secure the change which we ask for, we act only as business men, and that we stand ready to accept any fair compromise between the extremes of party, which will insure stability to the iron interest, and fair remuneration to the labor and capital engaged in it.

THOMAS CHAMBERS,
President.

Cost of making a ton of Pig Iron on the Schuylkill River, at the Furnaces situated between Norristown and Spring Mills, in 1849.

DATA.

2⅔ tons of iron ore delivered at furnace,....	at $2 00 per ton,		$5 33
2 " of coal in the furnace, / ¼ " extra for steam, smith fires, etc.,	2¼ at 3 25 "		7 31
1 " limestone, ...	at 75 "		75
			13 39
Furnace labor, ...		2 00	
Other expenses, labor, wear and tear, superintendence, interest, etc., ...		2 11	4 11
			$17 50

IRON ORE.

	Labor.	Other items.			
Ore leave,			40 cts. per ton, at 2⅔ tons,	1 06	
Mining,	$1 00		"		2 67
Hauling,	50		"		1 33
Weighing, etc.	10		"		27
	1 60 +	40		1 06 +	4 27 = 5 33

COAL.

	Labor.	Other items.			
Rent,..............		35 cts.	at $2\frac{1}{4}$ tons,	$78\frac{3}{4}$	
Operators' profit,.....		18 "	"	$40\frac{1}{2}$	
Mining, ...	90 cts.		"		2 $02\frac{1}{2}$
Charges on lateral roads, at 25 cts.,.........	13 "	12 "	"	27	$29\frac{1}{4}$
Wear and tear, 15 cts.,	12 "	3 "	"	$6\frac{3}{4}$	27
Incidental labor,......	7 "		"		$15\frac{3}{4}$
Reading railroad, 1 30,	70 "	60 "	"	1 25	1 $57\frac{1}{2}$
Interest.............		5 "	"	$11\frac{1}{4}$	
	1 92	1 33		2 99 +	4 32 = 7 31

LIMESTONE.

	Labor.	Other items.			
Quarry leave.........		10 cts. at 1 ton, 10			
Quarrying, ...	25			25	
Hauling, ...	40			40	
	65	10	10 +	65 =	75

GENERAL EXPENSES.

	Labor.	Other items.	
Furnace labor, ...	2 00		
Other labor and expenses,................	1 11 +	1 00	
Total general expenses, ...	3 11 +	1 00 =	$4 11
Forward Ore,..........................	4 27 +	1 06 =	5 33
" Coal, ...	4 32 +	2 99 =	7 31
" Limestone, ...	65 +	10 =	75
	12 35 +	5 15 =	17 50

The above shows the amount of labor represented by a ton of pig iron to be $12 35, and the amount of other items, not labor, composed of profits paid for the privilege of mining the minerals, coal operators' profits, and interest on investments, etc., to be $5 15.

Every anthracite furnace thus situated, making 5,000 tons of metal per annum, will benefit the following interests thus :

Clear profits to the Reading and other lateral railroads, for transporting coal to furnace, ...	$1 62	$8,100
Rent to owners of coal lands, ...	78½	3,925
Profit to the coal operators,.................	40½	2,025
Ore leave to owners of ore lands,.............	1 06	5,300
Quarry leave to owners of limestone quarries,..	10	500
Capitalists and storekeepers, ...	1 18	5,900
	5 15	25,750
Laborers engaged in mining and transporting, etc., and about the works, ...	12 35	61,750
	17 50	87,500
Transportation to market, 75 ; and drayage, 50,	1 25	
Selling and guarantee commission, 5 per cent.,..	1 00	12,500
Storage, weighing, etc., ...	25	
5,000 tons pig iron cost when sold, ...	20 00	100,000
5,000 tons sold at present market price, ...	20 00	100,000
Profit to the manufacturer,..................	00	00

Cost of Merchant Bar Iron manufactured on the Schuylkill River, from Pig Iron costing $17 50 *at the Mill.*

DATA.

1⅓ tons of pig iron, at $17 50,.........................	$23 38
2¼ tons of coal at 3 25,.........................	7 31
Labor per ton, ...	15 00
Interest, $1 00 ; wear and tear, $1 00 ; general expenses, $1 31,..	3 31
	$49 00

INTERESTS BENEFITED.

	By Furnace.	By Rolling Mill.	Total.
Stockholders of Reading and lateral roads, clear profits for carrying coal, in making $1\frac{1}{3}$ tons pig iron, and in making one ton of bars, ...	2 16	1 62	3 78
Rent to owners of coal lands,........	1 $04\frac{1}{2}$	79	1 $83\frac{1}{2}$
Profit to coal operators,...........	54	41	95
Ore leave to owners of ore lands,....	1 $41\frac{1}{3}$		1 $41\frac{1}{3}$
Quarry leave to owners of limestone lands,.........................	$13\frac{1}{3}$		$13\frac{1}{3}$
Capitalists and storekeepers,........	1 $57\frac{1}{3}$	2 31	3 $88\frac{1}{3}$
	6 87 +	5 13 =	12 00

LABOR.

Coal miners and laborers, ...	3 66	2 75	6 41
Labor for transporting coal,.........	2 10	1 $79\frac{1}{2}$	3 $89\frac{1}{2}$
Mining, hauling, etc.,...............	5 69		5 69
Quarrying and delivering limestone,..	87		87
Furnace labor, ...	4 14		4 14
Mill labor, \$15, and wear and tear, \$1,		16 00	16 00
	16 46 +	20 54 =	37 00
Brought down, ...			12 00
			49 00
Add transportation to market, ...		\$1 00	
Porterage, etc.,............................		50	
Com. and guarantee, 5 per cent. on \$55, ...		2 75	
Charges, ...		50	
			4 75
			53 75
Cr. by sales at \$55 per ton,...........................			55 00
Profit to manufacturer, ...			\$ 1 25

Every complete establishment producing 10,000 tons of merchant bars per annum, pays to diversified labor, at	$37 00	per ton,	$370,000
To the owners of coal in the ground, at..	1 83½	"	18,350
To the coal operator, clear profit, at.....	95	"	9,500
To the owners of ore lands, for ore leave, at	1 41	"	14,100
To the owners of limestone quarries, for quarry leave, at	13⅓	"	1,333
To capitalists, for the use of money, interest, etc., at........................	1 50	"	15,000
To Railroad and Canal companies, clear profits over and above expenses, for carrying coal to works, at	3 78	"	37,800
To storekeepers and others, for merchandise, etc., oil, brass, fire-brick, etc., at..	2 39⅙	"	23,917
Cost at works, at	49 00	"	490,000
Transportation to market, say at	1 00	"	10,000
Drayage and hauling, at	50	"	5,000
Storage and other expenses, at	50	"	5,000
Commission for selling and guarantee, at 5 per cent., ...	2 75	"	27,500
	53 75	"	537,500
Cr. by sales at market price, at.........	55 00	"	550,000
Profit to manufacturer, at	1 25	"	12,500

Such is the catalogue of interests benefited by the manufacturer of iron; among which he distributes $537,500 in making 10,000 tons of bars, whether he derives any profit from his business or not. Is he not, therefore, emphatically wasting his time, money, and experience for the benefit of the public? To do this amount of business, his capital invested in buildings, machinery, and ready cash, must amount to $500,000; for which, on the supposition that he can sell his iron for $55 per ton, which is above the present market rates, he realizes about 2½ per cent. per annum. Such is the compensation he now receives, if indeed any, for keeping in motion a branch of industry which employs more labor, and disburses more money through the ramified channels of trade, than any other—actually creating capital which otherwise would never exist.

When his works are in operation, he supports, directly and indirectly, more than 1,200 men, with their families, in all, not less than 6,000 people; whose aggregate earnings we have shown to amount to $370,000 per annum. When such an establishment is closed by British competition, the means of living to these 6,000 people are suddenly cut off. Does any one, then, seriously assert that the laboring man and his family are not benefited by protection? The advocates of free trade, nevertheless, do not cease to tell them, either ignorantly or designedly, that it is the manufacturers only, as a class, who are benefited.*

* REMARKS SUBMITTED BY THE AUTHOR OF THE MEMORIAL.

A PROTECTIVE MARKET.

There is great misapprehension on the subject of the PROTECTION asked for industry. The term is ill-chosen; because it implies special favor granted to particular branches of manufacture. But it is far from being a mere concern of individuals; it can be shown to be equally a matter of public policy. Suppose the makers of iron in Great Britain and the United States to have equal advantages, and that the manufacture is carried to the utmost extent, and the lowest point of remuneration by the home competition in each country; and that the average price in each is the same. If any state of industry could, with advantage, dispense with protection, it would be the case supposed. But it would be clearly the interest of both countries to protect their home markets, even in this case of perfect equality. It would be sound policy to keep the manufacture of an article so important as iron in prime vigor and progress, that the quantity might be increased, and the price reduced by the gradual process of home competition, which, in a protected market, is a severe but sure operation. One of the greatest trials the manufacturers encounter in such cases is, that fluctuation

The owners of coal lands in Schuylkill county, many of whom live out of the region, and are engaged in other occupations, derive an income of $1 83 per ton, on every ton of bar iron made ; but when the works cease operating, this income is at an end.

which occurs every few years in all mercantile communities. To bear up under all these, and maintain the full vigor of protection, is a hard trial upon makers of iron, the more so, as in their case their expenses do not admit of being abridged, nor can their manufacture be diminished under a limited demand without heavy loss. Seasons of depression must come in both countries in the case supposed—periods when the markets of each would reject and be unable to consume the ordinary quantity. It must be thrown somewhere, for neither makers nor merchants are able nor willing to hold the surplus of iron until business recovers its tone and makes its usual demand. If the iron thus remaining on hand in Great Britain is thrown into our markets, it will wholly break them down if firm, and increase and continue the depression if already down. Between the two countries, while prices were up, there would be no transactions in iron, but the conflict would be incessant in periods of depression. It may be safely assumed, however, that without help the manufacturers could never recover from such a conflict; a few years would end the struggle by prostrating a large portion of those engaged. The business would have to be reorganized.

It would be sound policy, therefore, to protect each of these markets from the irregularities of the other. This course is best for the makers of iron, as well as for those who are special consumers, to whom it insures, in the long run, the cheapest supply.

FOREIGN IRON—ITS INFLUENCE ON PRICES.

As every country is dependent mainly on its own industry for its supplies, it is important that the industry which furnishes these supplies should be suitably sustained. The prices of the nine-tenths furnished at home should range at such rates as to keep the production active and increasing. Unless it can be demonstrated that the whole supply could be permanently imported cheaper, it would be suicidal to extinguish the industry on which we are dependent for nine-tenths, in a vain experiment to purchase cheaper elsewhere. The prices in the home market should be such as are made by fair competition in the home market, in which all parties interested can take care of themselves. If our iron is made at home, all the labor which goes into the cost should be adequately compensated; the farmer who furnishes food for man and horse, the manufacturer who furnishes raiment, the laborer and operative who are immediately employed in the production—all these and the consumers must settle the price; the elements are among them, and their combined action must maintain a result the nearest to justice, because they all look to their own interests.

It is unjust and unwise to disturb and change this result by introducing a new

The coal operator, also, has a direct profit in every ton of bar iron made, to the extent of near one dollar per ton. The owners of ore lands and limestone quarries, who are in most instances farmers, may cultivate their fields, and raise vegetables for supplying the

element in a supply derived without restriction or regulation from foreign trade. Upon that trade we are not in any sense dependent for pig and bar iron; we should, indeed, at this time, be makers and consumers of a much larger quantity than we have yet used, if we had not imported a ton of iron the last twenty years. We make the *quantity* we consume much cheaper than we could import it. Is it just that the tenth of our consumption which we import should regulate, to the injury of the makers of the other nine-tenths, the prices of iron in this country? Yet the price is for the most part controlled by the movements of foreign trade. It happens that our seaports are also the chief markets for distribution of our domestic iron. The prices of every country or district are made at its chief markets. If the consumption of iron on the seaboard is 300,000 tons per annum, the import of 50,000 of foreign iron will control the prices, because it comes in to be sold for what it will bring. It is at once offered below the domestic article, and consumers, seeing a disturbing cause in the market, pause until the effect is seen. A pause in the purchase of iron produces a fall, because some sellers must realize, and buyers take the advantage and keep it. The whole mass of the domestic iron is brought to market to keep pace with consumption, and the price demanded is a remunerating rate, and unless this is obtained the business must perish. The quantity imported is a mere overplus—a remnant from British markets, the sale of which at high or low rates is not a very important or vital matter to the manufacturers who sent it. At most, it is but 10 per cent. of their product, and may be considered as their profit, greater or less, as sold. What is vital to them is their *home price;* if that is fair on the average, they can afford to risk 10 per cent. of their production in our market. Let any one who knows how prices are made to vary, not only by actual events, but by rumors and suspicions, reflect upon the effect of an additional 10 per cent. of a foreign article thrown in upon a previously balanced market, and he will perceive not only the necessary depression, but the injustice of it to the industry affected.

But there is a feature in foreign trade which greatly increases the mischief of leaving the market under its control. It is, to a most extraordinary degree, uncertain and fluctuating. The importing merchants are governed in some degree, doubtless, by the actual demand, and their imports might, if exhibited separately, show some regularity. But a large portion of the imports are sent upon speculation, and the quantity depends on markets abroad and the thousand contingencies which may determine a larger or less export to our shores. The irregularity of our imports of iron from Great Britain is so striking as to demonstrate the impolicy and injustice of making the prices of the domestic product subservient to it. Beginning with the year 1820, coming down to 1845, and

wants of the workmen, at the highest market rates, without the necessity of losing time and money in going to the city, and at the same time receive $1 56, from the iron manufacturer, on every ton of bars made, for the privilege alone of digging the ore and

leaving out the fractional hundreds, the following figures exhibit the number of thousands of tons of iron of all kinds imported into the United States from Great Britain, each year in its order:

Year.	Tons.	Year.	Tons.	Year.	Tons.
1820,.....	8,000	1830,.....	21,000	1840,.....	72,000
1821,.....	9,000	1831,.....	41,000	1841,.....	112,000
1822,.....	15,000	1832,.....	45,000	1842,.....	107,000
1823,.....	13,000	1833,.....	62,000	1843,.....	38,000
1824,.....	11,000	1834,.....	47,000	1844,.....	102,000
1825,.....	13,000	1835,.....	63,000	1845,.....	68,000
1826,.....	12,000	1836,.....	91,000	1846,.....	—
1827,.....	21,000	1837,.....	54,000	1847,.....	—
1828,.....	22,000	1838,.....	78,000	1848,.....	—
1829,.....	17,000	1839,.....	85,000	1849,.....	315,000

These figures show a variation in the supply of iron derived from Great Britain of from 10 to upwards of 200 per cent., between one year and the next. Small as this quantity appears, compared with our whole consumption, it would always control the prices in New York, and thence those in the country. How little these fits and starts of commerce are like the sober pursuits of industry at home, where the annual product only varies to increase with the gradual increase of labor, capital, and consumption!

How can it be just to make the laborer's wages depend upon the variable movements of foreign trade?

MANUFACTURE OF IRON.

We cannot manufacture iron in the United States as cheaply as it is made in Great Britain. Because:

1. The erection of furnaces and machinery costs from a third to a half less there. Iron enters largely into these constructions. The wages of all mechanics are not more than half our rates.

2. The wages of the operators in iron works are about half that which is paid here.

3. The cheapness of money there has enabled the manufacturers to prosecute their business with more vigor and more economy; and to introduce at once many modes of saving expense, the cost of which cannot be reached by makers here.

quarrying the limestone, which, but for the iron works, would be utterly worthless. These advantages, however, result only when the works are in blast; for, when Free Trade puts out the fires, neither ore nor limestone are wanted, nor can the farmers sell their

4. The manufacture in its present improved processes has many years the start there. We might, and, with our home market secured to us, could soon equal them in all departments, as we do in many now.

If we could import and pay for our whole supply at the low rates, it would be more difficult to refute the advocates of the cheap market. The iron market in Great Britain affords a spectacle of fluctuation and speculation which has scarce any parallel. The range of prices varies from below the cost of manufacture to 150 per cent. advance upon the actual cost. It has been with British manufactures, for the last twenty years, a constantly recurring feast or famine. Iron being an article not subject to deterioration, it is deemed safe to hold, and speculators step in when prices are at the lowest. Not only so, but at extremely low rates, iron enters into a large consumption for which it is too expensive at higher rates. At the low rates, this increased consumption begins; contracts are made, enterprises are commenced, plans and estimates are gone into, which produce, at last, an effective demand for iron sufficient to enable makers and holders to advance the rates in proportion to the wants of buyers, who hasten to supply themselves when the advance begins.

These fluctuations in price cannot be wondered at if we note the progress of the manufacture. The following is the estimate of the quantity produced in the years named:

Year.	Tons.	Exported to all the world. Tons.	Consumed.
1810, ...	294,642	—	—
1820, ...	368,000	91,766	276,234
1825, ...	581,367	69,328	—
1830, ...	678,417	130,417	—
1835, ...	1,000,000	219,203	—
1840, ...	1,500,000	268,328	1,231,672

The increase in product in 30 years was over 500 per cent.* The increase in the export was less than 300 per cent. The increased consumption was very nearly 500 per cent. There was in that period no parallel to this progress in the world. It is not hard to comprehend the share which this increased consumption of iron had in the *material progress* of Great Britain.

If this manufacture is not overdone in Great Britain, it would be difficult to

* The increase in 50 years was 100 per cent.

grain and vegetables to the workmen, whose ability to purchase departs with their occupation.

The clear profit to the railroad and canal companies in carrying the one article of coal from the mines to the works, is $3 78 per ton of bars made, and when

find any business overdone. It is only possible to keep up this large production through the operation of those fluctuations, by which, in times of depression, the iron is taken into a large consumption at the low rates, and by which the makers indemnify themselves for the losses of one period by the high prices of another. By this system, they export largely when they can do no better, and raise the prices so high at times, that exports must be greatly diminished. Upon this system of variation, the makers there thrive; but can these fluctuations be introduced elsewhere with advantage, or even without ruin? They are an incident to over-production there—they are an incalculable injury when brought to bear upon our industry.

PROTECTION OF PRICES.

As the progress of the manufacture of iron in Great Britain is one of the greatest achievements of industry in modern times, it may be worth while to consider what protection fostered and secured this wonderful growth.

Previous to this growth, pig iron had been scarcely known in commerce. England had no rival in the production of that article that could disturb her markets. The price may be shown thus:

1782 to 1803	20 years,......................	£3 00 to £9 00	Average £6 00
1803 to 1818	15 years,......................	£7 00 to £9 00	Average £8 00
1818 to 1840	22 years,......................	£4 15 to £11 00	Average £6 00
1840 to 1849	9 years,......................	£2 00 to £5 00	Average £3 5

It will be seen that down to the year 1840 from 1782, in which period the production of iron in Great Britain had increased from 150,000 tons to 1,500,000 or tenfold, the makers enjoyed a price averaging over £6, and very seldom below £5. Upon this price, the business flourished beyond precedent. After 1840, it became evident that there was an over production, and the British market broke down. The price of Scotch pig, which, after 1840, controlled the market, fell to £2, and even below that rate. It became the subject of speculation and fluctuation beyond any article of commerce.

It was the protection of this continued high price, for 50 years which stimu-

such an establishment, situated on the line of these improvements, stops for a year, the loss to the stockholders and bondholders of these improvements amounts to $37,800 per annum. Besides coal, there is an additional profit in transporing other raw mate-

lated the production of pig iron. The revenue duty of 27½ per cent., was of no consequence; the importations of pig iron were too small to have any effect upon the markets.

Large importations of bar iron were made into Great Britain between 1782 and 1840; but that it may be clearly seen how far this importation interfered with the domestic product, note the prices of Russian and Swedish bars, the only kinds largely imported.

	Years	Period		Prices	Average
Russian bars.—	1782–1795	13 years,........	from	£10 15 – £16 00	Average £16 00
"	1795–1803	8 years,........	from	£17 00 – £25 00	Average £21 00
"	1803–1820	17 years,........	from	£12 10 – £22 00	Average £16 00
"	1820–1840	20 years,........	from	£14 00 – £26 00	Average £19 00
"	1840–1849	9 years,............		——	Average £16 00
Swedish bars.—	1782–1795	....	from	£14 15 – £18 00	Average £16 00
"	1795–1803	....	from	£19 00 – £26 00	Average £22 10
"	1803–1820	....	from	£15 10 – £21 00	Average £17 00
Since 1820 the average has been about					£13 00

The above prices are exclusive of duty, which was increased from £2 16*s.* 6*d.*, in 1782, by *ten different advances*, to £6 10*s.* (and £7 18*s.* 6*d.*, in foreign ships), in 1820.

These prices were, indeed, in no degree an obstacle to the British manufacturers; they were, in fact, so high that no heavy importations could take place.

The largest importation in any year between 1800 and 1814 of bar iron, was 52,873 tons in 1802; during the nine years of that period the importations did not reach 30,000, and was sometimes below 20,000 tons.

From 1815 to 1840, the largest quantity of bar iron imported into Great Britain was 25,033 tons, in 1836, but the average for the whole period was considerably below 20,000 tons.

The British manufacturers required no legislative aid after 1800, yet such caution was used that the duties were increased from time to time to 1820, and were not removed until 1825.

rials, miscellaneous articles, the manufactured iron, and passengers to and from the works.

Again, the capitalist and owners of bank stock, who

PROTECTION IN UNITED STATES BY PRICES.

The British maker had competitors who furnished iron exclusive of duty at from $65 to $100 per ton.

The competitor of our manufacturers has furnished bar iron

From 1815 to 1830 at $50	Highest	$72 00
	Lowest	31 00
From 1830 to 1849 at $38	Highest	55 00
	Lowest	22 00

But low as the averages are, compared with those against which the British manufacturer had to contend, they afford an inadequate idea of the destructive effect of this competition. The averages are comparatively high from the excessive range of the fluctuation. For three years, from June, 1820, to July, 1824, the price did not exceed $46, and did not average over $42.

From March, 1827, to July, 1836, the price did not exceed $46.

From April, 1829, to October, 1835, the price did not exceed $37.

From March, 1830, to March, 1833, the price did not exceed $33, and for a year of this period it was under $27.

For three years, including 1841 to 1843, the price was at $24, and during 1842 as low as $22.

It is such prolonged depressions as these which seriously injure, if they do not ruin, the maker of iron in the United States. He cannot meet such exigencies, either by reduction of his expenditures, by reducing wages, or by diminishing the amount of his product. He must continue his business at a serious loss for years, or he must stop and be ruined.

It cannot be doubted that these periods of low prices have hindered the progress of this branch of industry to a very important extent. It is scarcely extravagant to say that with the same comparative protection which has been enjoyed in Great Britain, the product here would now have been scarce less than that of that country. If the home market had been equally secure to the makers here, as that of Great Britain was to the makers there, the consumption of iron here might now be 1,500,000 tons. The higher price would have been no obstacle, for where all labor receives a corresponding compensation, the price is no obstacle in the exchange of labor. These fluctuations in prices, introduced from Great Britain, have proved an incalculable evil to the whole industry of the country. The fact that a portion of our annual supply has been imported at a very low cost, much lower than it could be produced for here, is no alleviation. Instead of consuming more, we have consumed less. Our consumers say that they work up far less iron at the low rates than when business is proceeding on the basis of home prices.

If, owing to the low prices of British iron, we consume 200,000 tons less of

discount the business paper of the manufacturer, pocket $1 50 for every ton of bars made.

Besides the above-named, the draymen and the

domestic iron, we prevent the circulation of a value of $10,000,000, which, at $50 as the average per ton of pig and bar, would be its cost in food, labor, clothing, etc. All who are concerned in this great exchange are thrown out of their usual routine of employment. The farmer loses his market—the laborer his wages—the manufacturer his living—all are made less able to consume, and of course others who are dependent on them feel and suffer by the change, until the $10,000,000, by endless ramifications of the channels in which its benefits would have been felt, becomes hundreds of millions in its consequences. The effect of stopping the domestic manufacture is to throw business entirely out of the usual channels. It is too absurd to be held by any one, that the elements which would go to make 200,000 tons of iron in the United States, could be made available to import that quantity and pay for it. If imported, it must be paid for in something else than *iron ore*, *coal*, *wood*, *veal*, *mutton*, *potatoes*, *turnips*, *oats*, *rye*, *corn*, *and American labor.*

FREE TRADE.

It is a strange feature of Free Trade doctrines, that while they regard commerce as the great patron and regulator of industry, they wholly omit to make any allowance for the effect of commercial movement upon prices. The ability to make goods cheaply implies, with them, a willingness to sell them cheaply, without change of price. If all taxes, duties, and restraints were removed from commerce, the advocates of free trade seem to think the very facility of movement and transportation would furnish all the stimulus industry requires. They omit all notice of the fact that merchants, to whose tender mercies the producing classes are invited to commit themselves, are as much addicted to habits of thrift as other people; and that prices fluctuate more by these movements, and are more influenced by their operations, than by the efforts of producers. Merchants, in proportion to their number, are far better paid than manufacturers, and when the latter are starving, the former are often making large profits. It is quite as acceptable to the merchant to make a large profit on a few goods, as a small one upon many. The interest of the producing classes is to furnish a large product and a large exchange, that the comforts, luxuries, and benefits of mutual industry may be extended to the largest number. It is their interest to reduce commercial power and influence to its just minimum, because it is a charge upon industry. All the profits of the merchant are laid upon the consumer, and proportionably reduce the ability of the consumer to increase his consumption, and thereby make a large demand upon the producer. The truth is, that merchants have grasped a power and wield it for their own benefit, which enables them to oppress both producer and consumer. However necessary this may be, the merchants are certainly not the appropriate patrons of industry; they, as shrewd men usually do, take care of themselves, and whether

8

laborers along the wharves derive a benefit, and last of all, the commission merchant is sure to secure a handsome living out of the enterprise of the manufacturer.

All these vast interests are secured in their profits,

trade is free or not, no men can be in a better position to keep guard over their own interests. They survey the whole field of trade; occupying an intermediate post between consumer and producer, it is their interest to buy as cheaply as possible and sell as dearly. It is at their instigation that the doctrines of free trade are so loudly proclaimed. They and their friends are no doubt sincere. There is one great fact, however, which cannot be explained upon their principles—that industry, widely diffused industry, manufacturing industry upon the mighty scale now seen in Europe and the United States, has grown up under the protection of commercial restrictions. The productive powers of man were never exhibited before the days of the protective duties as they have been since. Commerce was free when Tyre, Carthage, the Grecian cities, Alexandria, Venice, Genoa, the Hanse towns, Holland, at their several periods of prosperity usurped the trade of the world. The merchants were then all in all—the merchants were princes, the producers slaves. Free trade would rapidly tend to the same results now.

If the question related solely to the prosperity of trade, there might be force in the position, that men in trade should be allowed to take care of themselves. The problem for solution is not what will most promote the interests of those engaged in commerce, but what will best promote the interests of all.

So far from being that department of industry which most requires the care of government, or rather best deserves to have its wishes granted, commerce is really a tax, an incumbrance upon industry, to be reduced as far as practicable. It is the expense incurred in distribution—an expense which, like all other mere expenses, should be kept at the lowest point consistent with the end in view.

What, then, will most promote the comfort and material well-being of the mass of the people? We reply, that industry which furnishes the largest product for the consumption of the masses, whilst all the producers receive, by a fair exchange for their labor, their full share of those articles which minister to comfort and physical well-being—that industry which, while it is thus successful in securing physical benefits, has a surplus large enough to maintain a good and efficient government, and all the advantages and enjoyments which belong to education, morality, philanthropy, and religion. Are these benefits to be obtained by merely removing restraints from the plans and movements of men in trade? This is no more true nor wise than to say that there should be no *checks or locks* upon the wheels which carry the goods to market.

We must inquire what are the circumstances in which men will and can exert their utmost productive powers. Clearly, where they consider their compensation most secure, and where the division of labor can at once, and in the same

before the manufacturer can receive a dollar for himself. Are they not all, then, as much interested in the success of the business as he?

Not less than forty millions of dollars are paid out every year in this manner to the public, through the

locality, be carried to such a point that a large exchange of labor can be effected, free from the expenses of transportation and intervening profits. Producers being consumers, and these being also producers, that is their best business which is done most directly, because it is that in which they are most likely to obtain mutual compensation in the results of their mutual labor. In all large manufacturing operations, one of the first considerations entertained by those who contemplate such undertakings is, the facility of selling the products in such quantities as may enable the manufacturer to sustain his business at the low prices to which competition may force his sales. The ready sale of the product is vital to the business, even if the profits are reduced to zero. The annual expenses of a blast furnace are from $200 to $300 each day—an expenditure which will rapidly absorb the working capital of its owner, if not replenished by corresponding sales. In many of the iron establishments of this country, the daily outlay is equal to, and in some over $2,000. A regular market is, therefore, indispensable. The manufacturer may estimate with reasonable correctness the demand and the competition of the home market, but must be wholly at fault in conjecturing what interference may come from abroad. He knows that his market at home may be disturbed, and, for a time, destroyed by any undue ingress of the foreign article; for experience has taught him that his sales are at an end for a time, when a cheaper article appears. When this happens, buyers become "bears, and operate for a decline," and this they effect by ceasing further purchases, until the market finds its lowest point—that is, until the necessities of the manufacturer compel him to come to the terms of the buyers.

NECESSITY OF A HOME MARKET.

The necessity of securing the home market to the home producer may be thus stated. A manufacturer has observed that his country has, for a long period, been supplied with a certain article at a range of prices at which he thinks he could furnish it. He consults the consumers of the article, and they encourage him to go on, agreeing to give him the preference. He makes a large outlay, and begins his work. His goods go off freely, and he has the market. The foreign article must now be withdrawn or reduced in price; it cannot be withdrawn, for there is no other market, and the price is reduced. The home producer is now receiving his first lesson, and he must reduce his rates also, at the first stage of his operations; the foreigner only after a long period of success. It becomes now a struggle for existence, the foreigner having the accumulated wealth of a long career, determines to extinguish his young rival, and again reduces the price. The home producer

operations of all the iron works in the United States, when the business is prosperous. With what truth can it be said, then, that protection is a tax upon the many for the benefit of the few? It is true that iron, at the present time, is furnished at low prices by Great Britain, but let the home manufacture be de-

applies to the government for protection, and though the whole array of free trade arguments are brought to bear against the application, it prevails, and a specific duty is laid which gives the producer a price at which he can maintain his production in full vigor. The foreigner, determined still to conquer the market, reduces his price according to the duty, or in other words, pays the duty and enters the lists again. At this stage of the struggle, the whole quiver of free trade weapons are let loose upon the monopolist, who is charged with receiving a premium to the whole amount of the duty, to sustain a manufacture that ought never to have been started. The absurdity of not buying altogether in this cheap market, strikes the philosophers of the closet so strongly, that they cannot express their surprise at the dull intellects of mere men of business. To resume our case: the home producer is again forced to ask further protection, and to say that his business must perish if he does not obtain it. Again common sense prevails over theory; a heavier duty is laid, and his business revives, though suffering severely from these interruptions, and by no means in the state of efficiency it would have been but for their influence. If it be supposed that the foreigner is unable to continue the struggle unaided, he applies next to his government for the removal of certain taxes, charges, duties, which bear upon his products, for the avowed object of enabling him to retain the foreign market, of which he is deprived.

Thus may a struggle be carried on for many years, to the serious injury of both parties—perhaps to their ruin. The operatives engaged in the home product, thus injured, must suffer severely; while the advocates of free trade cry out "let them fight it out," the merchants, who are the real purchasers, find their interest greatly promoted by the contest, as the lion's share falls to them.

It is thus, too, that the cheapness of a foreign market, which makes it the very climax of free trade arguments, is caused by want of demand for its goods; that want of demand arises from home production, which deprives the cheap market of their customers. The more the foreign market is thus cheapened by home production, the more the necessity is increased to afford protection to that market, on which the home production is mainly dependent. Individual merchants and consumers are always prompt enough to avail themselves of a cheap market when it offers; but nations should never commit their people to the absurdity of relying upon any market because it is cheap. The policy of a nation cannot be changed with the productions of a market, but it is the business of individuals to watch the market and operate wherever advantage calls them. The United States cannot obtain her whole supply of iron from Great Britain in one quarter of the year, make it at home the next, and go abroad for

stroyed, and it will immediately be dear again. It cannot be doubted, that if all the iron works in the United States should be closed for only one year, and we should depend upon England for our entire supplies, British bar iron would bring $100 per ton in New York, in less than six months, and the country at the end of the year would be bankrupt, by the drain of $40,000,000 of specie, to pay for it.

it the next: nor can this be done if the quarters are extended to years, or to periods of five years.

It is worthy of remark, that while free trade theorists cry out *laissez faire*, as summing up all the wisdom needed by governments in the management of trade, they stop the mouths of laborers, artisans, and manufacturers as not knowing, or not to be trusted with their own interests. Who are to be "*let alone?*"—the merchants. These agents, these buyers, transporters, and sellers of the products of industry ask, by their friends of the free trade philosophy, to have the whole business committed to them—to be *let alone*—while they exert every faculty, every nerve, and all the shrewdness, superior knowledge, and address they possess, under the stimulus of all the selfishness of human nature in the prospect of gain, for their own benefit. But let it be noted that the producers and the manufacturer do not cry "*let us alone*"—their cry in Europe for the last century and more has been for *protection*, and it has been accorded. Under this protection the products of Europe have increased at a rate fivefold the increase of population. But this *let-us-alone* policy is put forth only in behalf of the foreign merchants; the merchant of the domestic products unites with his special patrons. The foreign merchant, whose business represents but a tithe of the business of the country, asks to have the interest of the nine-tenths committed to him. This cannot be denied; let the manifestos of free trade be examined, and it will be found that they mock at the manufacturer, scoff at his statements, complaints and petitions. It has been so for a century; yet the doctrine of protection has been in the main the policy of every modern civilized nation during that period, in which industry and the arts have made more progress in one century than in thousands of years before. It is true, that in England, and in some other countries, some departments of industry have been so long and so fully protected, that, with the advantage of cheap labor and cheap capital, they no longer require *protection;* and the persons interested in these branches of production may now join in the cry of free trade, as they naturally wish to remove all obstacles to carrying their goods to all the markets of the world. These exceptions only prove their rule; they are exceptions only because the rule existed. Let manufacturers, small and great, then, be heard, not only in their defence, but permit them to state what is required for their security and their success; let them be heard and regarded as representing the producing classes—the industry of the country.

"But," say the advocates of free trade, "not so, we would pay for it by the exportation of our breadstuffs."

Let us see.

Political economists usually allow, for each individual in this country, a consumption of the products of the land to the value of $50 per annum. The number of people supported by the iron business of the whole country, is *ab ovo usque ad malum* about 600,000, who, at $50 per head, consume annually $30,000,000 worth of breadstuffs. If these cannot find employment in manufacturing, they must become producers, and thus is a home market to this immense amount lost to the present producers.

The total exports of breadstuffs from this country to all the world, for the year ending December 24th, 1849, amounted to only $22,895,783, of which Great Britain and Ireland took $14,157,666, not equal in value to the iron we imported from that country last year. Thus it would appear that the iron business furnishes a larger market to the farmer at home, and at better prices than all the world beside. And furthermore, that Great Britain, which offers to make all our iron and everything else, takes less than the moiety of the farmers' products consumed by the iron manufacturer at his side.

Instead, then, of discouraging the progress of the home manufacture of iron, as the advocates of free trade propose to do, for the avowed benefit of the farmers, no policy having this tendency could be devised more injurious to the great agricultural interest. On the contrary, that policy would be the best for the producers of this country which would tend to double the present home manufacture of iron; in accomplishing which, they would at once make 600,000 new customers entirely dependent upon them for $30,000,000 of breadstuffs, which is more than the boasted market of the world took from them last year.

Every complete iron works, capable of producing 10,000 tons of bar iron per annum, supports, as we have said before, 6,000 people, and makes a market to the farmers in the sphere of its influence to the extent of $300,000 per annum. Suppose 1,000 farmers to participate equally in the advantages of this market, supplying beef, mutton, pork, butter, eggs, vegetables, etc., this will give to each ready sales to the extent of $300, one-half of which will be for perishable produce that will not bear exportation.

"But," says the opponent of protection, "it is unjust to tax the farmer for the iron used in making his hoes, harrows, ploughs, axes, etc."

Let us see what this tax amounts to among these 1,000 farmers. One thousand pounds per anuum is a large allowance for the average consumption of iron among farmers; this is sold to the blacksmiths, wheelwrights, and makers of hoes and harrows, by the manufacturers, at say three cents per pound, which is $30 for each. Beyond this, the farmer is not interested in the enhanced prices caused by a protective duty on iron, because the main cost of his agricultural implements is made up of the additional labor put upon them. Now, if the duty is added to the price, assuming it to be at the rate of $17 per ton, 1,000 lbs. would pay a duty of $7 60, which, deducted from $30, would, if sold at cost of importation, make the foreign article worth $22 40. This difference, then, of $7 50, is the utmost possible injury the farmer can sustain; while in compensation for this, if he can only supply all the workmen engaged in getting the materials, transporting, and manufacturing but 10 tons of bar iron per annum, he will sell them $300 worth of pork, butter, eggs, vegetables, and other products of the farm, besides making a market for his limestone, iron ore, lumber, and many other articles necessary in the construction and operation of an iron works. The loss to the farmer in time alone, not to speak of other expenses, in disposing of $300 worth of produce in a distant market, far exceeds the nominal tax he is supposed to pay for preserving a never-failing and invaluable home market, such as the iron manufacturer affords him.

If we resolve our importations of iron and manufactures of iron and steel into their original elements of cost, we shall find that in this shape we have been large importers of British breadstuffs, coal, iron ore, limestone, labor, etc.

Let us examine into this matter.

The imports into the United States from Great Britain for the fiscal year ending June 30, 1849, were as follows:

	Tons.	Value.
Bar iron,	173,473	$6,060,068
Hammered iron,	10,595	525,770
Hoops and sheets,	11,174	543,256
	195,242	7,129,094
Pig iron,	105,632	1,405,613
Steel,	6,690	1,227,138
Manufactures of iron and steel, etc.,	11,824	5,297,116
	319,391	15,058,961

COAL USED IN THE MANUFACTURE.

				Tons.
Of 195,246	tons of	bar iron, ham'd, etc., at 5	tons per ton,	976,225
Of 105,632	"	pig iron, at 2½..........	"	264,080
Of 6,690	"	steel, at 8..............	"	53,520
Of 11,824	"	manufactured, at 10......	"	118,824
Total coal imported in the form of iron,..............				1,412,649

IRON ORE USED.

	Tons.
For bars, ham'd iron hoops and sheets, at 3⅓ tons, ...	650,817
For pig iron, at 2⅔ tons, ...	281,686
For steel, at 5¾ tons, ...	38,468
For manufactures of, etc., at 7 tons, ...	82,768
Total ore imported in the form of iron, ...	1,053,739

LIMESTONE USED.

	Tons.
For bars, ham'd iron hoops and sheets, at 1⅓ tons,........	260,327
For pig iron, at 1 ton,................................	105,632
For steel, at 2⅛ tons,................................	14,216
For manufactures of, etc., at 2⅔, tons,..................	31,531
Total limestone imported in the form of iron,..........	411,706

BRITISH LABOR EMPLOYED.

Three-fourths of $15,058,961, ...	$11,294,221
Number of workmen earning say $200 per year (which is fully 25 per cent. above the average of wages in England), gives,................................	56,471
Persons supported, allowing 5 per head, gives.........	282,355

CONSUMPTION OF BREADSTUFFS.

In Great Britain the average consumption of breadstuffs per head may be $30* per annum, which will give us

* The average consumption of breadstuffs in the kingdom is set down by British writers at $38 per head, which would represent at that rate a total of $10,729,490.

the entire quantity of British breadstuffs imported in the form of iron for the last fiscal year, at.......... $8,470,650
The same number of people engaged in manufacturing a like quantity and description of iron in this country, would have consumed each $50 worth of the farmers' products, or.................................$14,117,750

which is almost identical in amount with our total exports of breadstuffs to Great Britain and Ireland for the last fiscal year.

But it is not the American agriculturists alone whose interests have been seriously prejudiced by a policy which discourages the home manufacture of iron, and permits the British manufacturer to supply us with $15,000,000 worth of iron per annum.

If we assume a locality in this country for the sake of estimating its value, where we may suppose this vast amount of iron and iron wares to be manufactured, we shall see that others who are interested in understanding the true policy of this country, have lost far more than they have gained, by depending on England for their iron.

Let us suppose this locality to have been on the Schuylkill River, near Philadelphia, than which none other could be selected more favorable to the English side of the question.

Profits to Reading and lateral roads, and Schuylkill Navigation, in transporting 1,412,649 tons of coal, at 72 cts.,	$1,017,107
Rent to owners of coal lands on do. at 35 cts., ...	494,427
Profits to coal operators on do. at 18 cts., ...	254,276
Ore leave to owners of ore lands on 1,053,739 tons iron ore, at 40 cts., ...	421,495
Carried forward, ...	$2,187,305

Amount brought forward, ...	$2,187,305
Quarry leave to owners of limestone quarries on 411,706 tons of limestone, at 10 cts., ...	41,170
Interest to owners of bank stock, profits to dealers in merchandise, oil, brass, etc., etc., belonging to the head of general expenses and interest, on 319,375 tons of manufactured iron, at $4, ...	1,277,500
Total loss to the above interests in one year, ...	$3,505,975

We need say nothing further of lost benefits, in carrying the manufactured goods to market; the passenger and merchandise traffic created by such an immense business and such a large dependent population; nothing of the enhanced value it would have added to all kinds of property; nothing of increased revenue to the State, arising from increased population and new taxables, as well as that which would accrue from increased tolls on her internal improvements; nothing of the immense advantage to the whole country of keeping $15,000,000 of money at home, to be used in developing the great resources of our own country in constructing railroads and canals to cheapen and facilitate intercommunication among our own people, instead of sending that much specie every year to England, the effect of which is to derange the regular business of the country, by distributing that great lever of trade—the currency. It would be impossible to calculate the loss which these and many other interests have sustained by discouraging the progress of the home manufacture of iron, under the delusive idea that it is the best policy to buy in the cheapest market; in which the money price, and not the labor price, is regarded as the criterion of cheapness.

STATEMENT

Showing the cost of making Anthracite Pig Iron in Wales, in 1849.

	£	s.	d.
2 tons of clay ironstone, at 10s.,	1	0	0
15 cwt. hematite ore, at 22s.,		16	6
2 tons coal in the furnace, at 5s.,		10	0
1½ ditto, steam, hot-blast, roasting ore, etc., at 5s.		7	6
10 cwt. limestone, at 3s.,		1	6
Wages,		9	0
General expenses,		6	0
Cost at the furnace per ton,	£3	10	6

Cost of making Coke Pig Iron in Wales.

		s.	d.
1 ton of clay ironstone,		10	0
1 ton of cinder,		5	0
15 cwt. of red hematite* ore from Whitehaven, at 22s. per ton,		16	6
3 tons of coal for coking, at 4s.,		12	0
17 cwt. of coal for the engine and hot-blast, at 2s.,		1	9
10 cwt. limestone,† at 2s.,		1	6
Carried forward,	£2	8	9

* As the price of this ore may be doubted by some persons, the items of cost are given as follows:

		s.	d.
The price on board the vessel at Whitehaven, reduced in August, 1849, from 12s. to		11	0
Freight from Whitehaven to Cardiff,		7	0
Railroad from Cardiff to Merthyr, 25 miles,		2	6
Loading and unloading railroad wagons,			6
Tram way from railroad to furnaces, loading and unloading tram wagons,		1	0
Cost at the furnace per ton,	£1	2	0

It is very generally used throughout Wales and Staffordshire, to mix with the clay ironstone of the coal measures.

† At Merthyr, the limestone costs about 1s. 6d. per ton, but along the valley above Newport it costs 4s. 6d. per ton. Three shillings is given as the average. In some places they use the blast furnace cinder for a flux instead of lime-

Amount brought forward,	£2	8	9
Wages,		6	0
General expenses,		6	0
Cost at the furnace per ton,	£3	0	9

STATEMENT

Of the cost of making Pig Iron in Scotland

		s.	*d.*
2 tons of raw coal, 4*s.*,		8	0
3½ tons of raw ore (equal to 1 ton 15 cwt. roasted), 5*s.*,		17	6
6 cwt. limestone, 7*s.*,		2	1
Coal for engine and hot-blast 1 ton, 2*s.*,		2	0
Labor at the furnace,		5	0
General expenses,		5	8
Cost at the furnace per ton,	£2	0	3

STATEMENT

Showing the cost of converting Pig Iron to Rails, in Wales.

			Per Ton.		
Assuming the cost of Pig Iron to be			£3	0	9
Refining—Fuel, 10 cwt. coke, at 9*s.*,	4*s.*	6*d.*			
Wages, refiner and helper, per ton		11			
Breaking and wheeling metal to forge,		1¾			
13 per cent. loss on pig, at £3 0 9,	7	10¾			
Cost of fining,	13	5½		13	5½
Cost of plate metal,			£3	14	3½
Puddling—15 cwt. coal to puddler, at 4*s.*,	3*s.*	0*d.*			
3 cwt. coal to engine, at 2*s.*,		8			
				3	8
Puddler and helper,	6	0			
Squeezer,		4			
Carried forward,	6	4	£3	17	11½

stone, because of the high price of the latter. No account is taken of that, as the loss in the quality of the iron is more than the gain by using the cinder.

			Per Ton.		
Amount brought forward, ...	6	4	£3	17	11½
Rolling puddled bar, ...		8			
1 extra boy at train, per day, ... 2*s*. 2*d*.					
2 ditto, dragging out, at 11½*d*., 1 11					
2 men weighing, ... 3 6					
1 man wheeling cinder, ... 2 1					
30)9 8 =		3⅚			
Average quantity rolled per day 30 tons.					
Ash fillers, ...		⅙			
				7	4
Loss 6 per cent. on plate, at £3 14 2½,				4	5½
Cost of puddling, ...	15	5½			
Cost of puddled bar per ton, ...			£4	9	8

The top and bottom of the rail is made from puddled bar re-heated and rolled, which costs as follows :

Fuel—12 cwt. coal to furnace, at 4*s*., ...	2	4¾			
3 cwt. coal to engine, at 2*s*., ...		8			
				3	0¾
Wages—Rolling per ton, ...	1	3½			
Heating "	1	8½			
				3	0
Loss 10 per cent. on puddled bar at £4 9 8,				8	11¼
Cost of making tops and bottoms, ...	15	0¼			
Cost of tops and bottoms per ton, ...			£5	4	8¼

A rail pile is ¼ tops and bottoms, at £5 4 8¼	£1	6	2
¾ puddled bar, at 4 9 8	3	7	3
Cost per ton of rail piles, ...	£4	13	5

Finishing Rails.

Fuel—12 cwt. coal to furnace, at 4*s*., ..	2*s*.	4¾*d*.			
3 cwt. coal to engine, at 2*s*., ...		8			
				3	0¾
Carried forward, ...			£4	16	5¾

					Per Ton.		
Amount brought forward, ...					£4	16	5¾
Wages—Cutting, wheeling, and piling iron,				6d.			
Rollerman, ...				5			
Roughing down, ...				4			
Catching, ...				3			
Hooking in, ...per day,	2s.	9d.					
Heave up roughing, .. "	2	3					
Ditto finishing, .. "	1	6					
Catcher ditto. .... "	2	0					
30)8 6 =				3½			
Heating, including helper, ...			1	8½			
1 extra helper to charge, ...	2s.	6d.					
1 ditto, coach, ...	2	6					
30)5 0 =				2			
Wages to heat and roll, ...						3	8
					£5	0	1¾
Sawing and hot straightening.							
1 man, "	2s.	9d.					
3 men, " 3s.,	9	0					
1 man, " 6s.,	6	0					
30 tons per day, ... 30)23 9 =				10d.			
Filing saws, ...				¾			
Cold straightening, ...				10			
Dressing, ...				4			
Patching, ...				1			
Inspecting, ...				1¼			
Total—Hot and cold straight'g and finishing,						2	3
Loss 10 per cent., at £4 13 5, ...						9	4
Cost of rolling and finishing, per ton, ..			18	3¾			
Carried forward, ...					£5	11	8¾.

	Per Ton.		
Amount brought forward, ...	£5	11	8¾
General expenses—such as, superintendence of mills, engineers, firemen, masons, blacksmiths, fire-bricks, oil, grease, fuel for smiths, iron and steel to mend tongs, heaters' and puddlers' tools, sand, cinder and ore to line and repair the furnaces, renewal of castings burned and broken, ...		6	0
Cost at the mill, ...	£5	17	8¾
Freight from Merthyr to Cardiff, ...		2	6
Cost of 1 ton rails at Cardiff, ...	£6	0	2¾

STATEMENT.

Being a summary of the preceding Statement, showing the cost of Fuel, Wages, etc., to the ton of Rails.

Pig at					£3	0	9
Fuel—Finery, 10 cwt. coke, at 9*s*.,........			4*s*.	6*d*.			
Puddling furnace, 15 cwt. coal, at 4*s*.,..			3	0			
Do., engine, 3 cwt. coal, at 2*s*.,...				8			
Tops and bottoms,							
12 cwt. coal, at 4*s*.,......	2	4¾					
Engine, 3 cwt., coal, 2*s*.,....		8					
4)3		0¾ =		9¼			
Rail finishing, furnace,							
12 cwt. coal, 4*s*.,.................			2	4¾			
2 cwt. coal, engine, at 2*s*., ...				8			
Total cost of fuel to the ton of rails,...						12	0
Wages—Finery, ...			1	0¾			
Puddling and rolling puddled bar,.....			7	4			
¼ wages tops and bottoms,..4)3 0 =				9			
Heating and rolling rails,............			3	8			
Straightening and finishing rails,......			2	3			
Total cost of labor to the ton of rails,.						15	0¾
Carried forward,.................					£4	7	9¾

	s.	d.	£	s.	d.
Amount brought forward, ...			£4	7	$9\frac{3}{4}$
This amount was reduced 10 per cent., on account of the selling price of rails going below cost—15*s.* $0\frac{3}{4}$*d.* less 10 per cent., or 1*s.* 6*d.* = 13*s.* $6\frac{3}{4}$*d.*, which is the present actual cost of labor per ton of rails finished.					
Losses in Manufacture,	*s.*	*d.*			
Finery, 13 per cent. on Pig at £3 0 9 =	7	$10\frac{3}{4}$			
Puddling, 6 per cent. on Plate at 3 14 $2\frac{1}{2}$ =	4	$5\frac{1}{2}$			
$\frac{1}{4}$ tops and bottoms 10 per cent. on puddled bar at, ... 4 9 8 =	2	$2\frac{3}{4}$			
Rails, 10 per cent. on rail piles, 4 13 5 =	9	4			
Total cost of losses, ...			1	3	11
General expenses, as before, ...				6	0
Cost of the mill, ...			£5	17	$8\frac{3}{4}$
Freight, ...				2	6
Cost of 1 ton rails at Cardiff, ...			£6	0	$2\frac{3}{4}$

STATEMENT showing difference in cost in English and American Labor in the Rolling Mill.

LABOR.	American price of labor, 1849, per ton.	English price of labor, 1848, per ton.	English price of labor less reduction of 10 per ct. 1849, per ton.
Puddler and his helper, ...	\$3 50	6*s.* 0*d.*	\$1 $29\frac{1}{2}$
Rolling and puddled bar, ...	$72\frac{3}{4}$	8	$14\frac{1}{2}$
Sundry labor, ...	$82\frac{1}{4}$	1 $8\frac{3}{4}$	$37\frac{1}{2}$
Shearing iron for piles, ...	21	6	11
Heater and his helper, ...	$87\frac{1}{2}$	1 $8\frac{1}{2}$	37
Rolling, ...	85	1 $11\frac{1}{2}$	42
Straightening and finishing, ...	1 $37\frac{1}{2}$	2 3	$48\frac{1}{2}$
Sundry labor, ...	1 $25\frac{1}{2}$	3	$5\frac{1}{2}$
American labor to ton iron, ...	\$9 $61\frac{1}{2}$		
English labor to ton iron, 1848, ..	....	15 $0\frac{3}{4}$	
English labor to ton iron, 1849, ..	....	....	\$3 $25\frac{1}{2}$

This does not show the *entire labor* in the rolling mill to the ton of iron, as in England they include engineers, overseers, firemen, masons, etc., etc., with materials, grease, oil, etc., all under the head of General Expenses.

To make the American account correspond, these items have been omitted :

They amount to	$1 38½
Add as above,..........................	9 61½
American cost of labor, ...	$11 00
And by proportion the English labor,.......	3 71

Or very nearly ⅓ the amount paid in this country.

Average and Comparative View of Prices of Pig Iron in Glasgow, for the last twenty years.

Year.	£	s.	d.	Year.	£	s.	d.	Year.	£	s.	d.
1830,....	5	0	0	1835,.....	4	10	·0	1840,.....	3	15	0
1831,....	4	10	0	1836,.....	6	15	0	1841,.....	3	0	0
1832,....	4	10	0	1837,.....	4	0	0	1842,.....	2	10	0
1833,....	4	0	0	1838,.....	4	0	0	1843,.....	2	16	0
1834,....	4	5	0	1839,.....	4	10	0				

Month.	1844.			1845.			1846.			1847.			1848.			1849.		
	£	s.	d.	£	s.	d.	£	s.	d.	£	s.	d.	£	s.	d.	£	s.	d.
January,..	2	0	0	3	5	0	4	0	0	3	13	4	2	7	8	2	7	0
February,.	2	5	0	3	14	0	3	17	6	3	13	4	2	9	8	2	11	7
March, ...	2	10	0	5	5	0	3	10	0	3	11	1	2	4	6	2	9	9
April, ...	3	5	0	5	7	6	3	6	0	3	10	8	2	1	8	2	8	0
May, ...	3	5	0	4	8	0	3	10	0	3	5	3	2	4	2	2	3	9
June, ...	3	5	0	3	5	0	3	8	0	3	5	0	2	3	0	2	4	4
July,.....	3	0	0	3	5	0	3	10	0	3	8	1	2	5	3	2	5	0
August,...	2	15	0	3	7	6	3	15	0	3	7	9	2	4	6	2	5	4
September,	2	10	0	4	2	0	3	13	6	3	6	0	2	4	10	2	4	0
October,..	2	12	6	4	10	0	3	9	6	2	19	10	2	2	9	2	2	10
November,	2	12	6	3	17	6	3	9	0	2	11	0	2	1	9	2	4	3
December,	2	17	6	3	16	0	3	12	6	2	7	6	2	2	0	2	7	2
Average,	2	14	0	4	0	3	3	11	9	3	5	0	2	4	4	2	6	1

Average price for the five years,	1840 to 1844, ...	59*s.* 2*d.*
" "	1845 to 1849, ...	61*s.* 5*d.*
" ten years,	1840 to 1849, ...	60*s.* 3*d.*

In the same year, the iron masters of Pennsylvania memorialized Congress; those of Maryland also held a meeting to examine into and report upon the cause of the depressed condition of the iron manufactures of that State. This report states, "that, previous to the passage of the tariff of 1846, there were in Maryland thirty-one furnaces and five rolling mills (of these, eleven furnaces and four rolling mills had been stopped), which produced, when in operation, 55,000 tons of pig iron, and 20,000 tons of bars and rails, per annum; the manufacture of which gave support to upwards of 50,000 persons; while in the coasting trade incident to it, a large number of men and vessels were employed in transporting a great proportion of the pig iron and rails to other States in the Union. That the chief cause of this extraordinary depression of the iron trade could be traced to the surplus production thrown upon the American market in 1848 and 1849 by the English manufacturer, glutting it, and thereby causing the stoppage of most of the iron works of that State. That the English and Scotch iron masters have perfect control over their labor, until it is reduced almost to the *point of subsistence*—while in the United States, the demand for labor is such that the iron masters cannot reduce wages below the price paid to laborers in other branches of business. At the regular quarterly meetings of iron masters in England, *the price of iron*, for the ensuing quarter is always declared, and the *price paid for labor* depends upon *the price of iron* so declared. The power which they possess over their labor in their ability to reduce the price, as the necessity of the case may require, and still continue to manufacture it. For example, during the years 1845, 1846, and 1847, the price of bar iron in Liverpool averaged respectively £9 4*s*., £9 13*s*., and £9 17*s*.; and before the close of 1848, the price was reduced to £4 15*s*., showing a reduction in less than one year of nearly fifty per cent. In Scotch pig iron, the average price in 1845 was £4 5*s*., and in 1848, £2 2*s*.

From the most reliable information, the cost of charcoal pig iron in the vicinity of Baltimore was from $22 00 to $23 50 per ton. The experience of the last four years has shown that the *ad valorem* duty, *without a minimum*, as laid by the tariff of 1846, has operated very injuriously to the interests of the American manufacturer. For, when the price of iron is high abroad, the duty is high at home, giving to the American manufacturer an incidental protection, which continues so long as the foreign market remains high; but as soon as the foreign market fluctuates, the duty falls with it; so that at the time when the highest duty is needed to enable the American manufacturer to sustain a competition with the foreign manufacturer, the protection is taken away—thus acting as a *sliding scale against* the American manufacturer. When the tariff act of 1846 was passed, the thirty per cent. duty on the price of iron at Liverpool ($50) was $15 per ton: the cost and duty added, made the price $65. But, for the last two years, the price has fallen from $50 to $27 per ton, and the duty from $15 to $8 per ton, making the cost of iron and duty $35 per ton, a fluctuation of $30 per ton. To sustain the American manufacturer he requires the *reverse* of the operation of the present *ad valorem* duty. When the price abroad is highest, he needs the least duty; and when it is lowest, he requires the highest. It is of the greatest importance to the prosperity of the American manufacturer that the fluctuations of the foreign market should have as little effect as possible upon the American. They may be lessened by a fixed specified duty on the part of the American government, or by a sliding scale of duties in favor of the American manufacturers—*not against them*, as the present *ad valorem* duty produces. Availing themselves of the low duties, the English manufacturers have sent large stocks of iron to the United States, which, from the very low rates of interest on capital at home, they can afford to hold until the regular wants for consumption absorb them."

In 1849, the depressing effect of the free-trade tariff on the manufacturing industry of the country was severely felt. The power to import was chiefly maintained by large remittances of railroad bonds and other evidences of debt in place of the specie which was daily diminishing in the vaults of the banks. The price of labor rapidly fell, and thousands of persons were thrown out of employment. A financial crisis was evidently approaching, when, fortunately for the advocates of free trade, the war with Mexico, and the acquisition of the gold mines of California gave a fresh stimulus to commerce and manufactures, which postponed for a few years the financial revulsion of 1857. But how different would have been the condition of the country, if the tariff of 1842 had remained in full operation a few years longer! In no part of the world had iron manufactures made more rapid progress, and in none were there so many improvements made in its production and manufacture. Furnaces were made to yield double the quantity, and the production would soon have exceeded the consumption; but, just at the time when iron manufactures needed most protection and were recovering from their depressed condition, the free trade theory of 1846 was adopted, which has ever since inflicted the most serious injury on the iron industry of the country, and retarded its growth fifty years.

Everything now, however, indicates that sooner or later the protective system will again raise its head in the United States, and become a settled policy. Whatever may have been the exertions of the English to diminish, or to temper the commercial revulsions in the United States, and however considerable the capital they send here to purchase public stocks and securities, the want of equilibrium, ever subsisting and continually increasing, between the value of exports and imports, which can never be reëstablished in that manner, and the consequent formidable revulsions, and their increasing violence, cannot fail to awaken every American to a full knowledge of the causes of the evil, and make him willing to apply the proper remedy.

The following tables, collected with great care from the official returns made to the Treasury Department, show the actual condition of the iron manufactures of the United States for 1850.

STATEMENT showing the population and value of the Iron Manufactures of the United States and Territories, for the year 1850.

STATES.	Population, 1850.	Manufactures of Pig Iron.	Manufactures of Iron, Castings.	Manufactures of Iron, Wrought.
Alabama, ...	771,623	$22,500	$271,126	$7,500
Arkansas, ...	209,897	—	—	—
California, ...	92,597	—	20,740	—
Columbia, Dis. of,	51,687	—	41,696	—
Connecticut, ...	370,792	415,600	981,400	847,196
Delaware, ...	91,532	—	267,462	38,200
Florida, ...	87,445	—	—	—
Georgia, ...	906,185	57,300	46,200	12,384
Illinois, ...	851,470	70,200	441,185	—
Indiana, ...	988,416	58,000	149,430	11,760
Iowa, ...	192,214	—	8,500	—
Kentucky, ...	982,405	604,037	744,316	299,700
Louisiana, ...	517,762	—	312,500	—
Maine, ...	583,169	36,616	265,000	—
Maryland, ...	583,034	1,056,400	685,000	771,431
Massachusetts, ..	994,514	295,123	2,235,635	3,908,952
Michigan, ...	397,654	21,000	279,697	—
Mississippi, ...	606,526	—	117,400	—
Missouri, ...	682,044	314,600	336,495	68,700
New Hampshire,	317,976	6,000	371,710	20,400
New Jersey, ...	489,555	560,544	686,430	1,079,576
New York, ...	3,097,395	597,920	5,921,980	3,758,547
North Carolina,.	869,039	12,500	12,867	331,914
Ohio, ...	1,980,329	1,255,850	3,069,350	127,849
Pennsylvania, ..	2,311,786	6,071,513	5,354,881	9,224,256
Rhode Island, ..	147,545	—	728,705	223,650
South Carolina,.	668,507	—	87,683	—
Tennessee, ...	1,002,717	676,100	264,325	670,618
Texas, ...	212,592	—	55,000	—
Vermont, ...	314,120	68,000	460,831	127,886
Virginia, ...	1,421,661	521,924	674,416	1,098,252
Wisconsin, ...	305,391	27,000	216,195	—
Territories. Minnesota, .	6,077	—	—	—
Territories. New Mexico,	61,547	—	—	—
Territories. Oregon, ...	13,294	—	—	—
Territories. Utah, ...	11,380	—	—	—
Total, ...	23,191,876	12,748,727	25,108,155	22,628,771

The census returns of 1850 show a deficit, under the free-trade tariff of 1846, in the manufacture of pig iron of nearly two hundred thousand tons, since it went into operation, which was supplied by importations from Great Britain.

The pig, bar, and wrought iron, steel and other iron manufactures imported into the United States, in 1850, amounted to $17,524,459 ; and the total amount of pig iron consumed, 1,042,929 tons.

Actual condition of the Iron Manufactures in the United States, in 1850.

PIG IRON.

Number of establishments in operation, 377 ; capital invested, 17,346,425 dollars. Materials used, and value :

Ore,	tons,	1,579,318	7,005,289 dollars.
Coal,	"	645,242	
Coke and Charcoal,	bush.,	54,165,236	

Number of persons employed, 20,298 ; average wages per month, 20 dollars 76 cts. Pig iron made, 563,755 tons; value, 12,748,727 dollars.

CASTINGS.

Number of establishments in operation, 1,391; capital invested, 17,416,361 dollars. Materials used, and value :

Pig iron,	tons,	345,553	10,346,265 dollars.
Old metal,	"	11,416	
Ore,	"	9,850	
Coal,	"	190,891	
Coke and Charcoal,	bush.,	2,413,750	

Number of persons employed, 23,541 ; average wages per month, 27 dollars 38 cts. Castings made, 322,745 tons; value, 25,108,155 dollars.

WROUGHT IRON.

Number of establishments in operation, 552 ; capital invested, 17,033,279 dollars. Materials used, and value :

Pig metal,..........tons,	251,491		
Blooms,............ "	33,344		
Ore,............... "	78,787	}	13,524,777 dollars.
Coal,.............. "	538,063		
Coke and Charcoal, bush.,	14,510,828		

Number of persons employed, 16,110; average wages per month, 25 dollars 41 cents. Wrought iron made, 278,044 tons; value, 22,628,771 dollars.

The total amount produced from the three descriptions of iron manufactures, in 1850, were as follows:

Pig iron,....................	$12,748,727	
Iron castings,................	25,108,155	
Wrought iron,................	22,628,771	
		$60,486,652
From which deduct the cost of raw materials, viz.:		
Pig iron,....................	7,005,298	
Iron castings,................	10,346,265	
Wrought Iron,...............	13,524,777	
		30,876,340
Total produce of iron manufactures,........		29,610,312

Comparison of the principal results of the census of 1840 and that of 1850:

	Census of 1840.	Census of 1850.
Number of blast furnaces,........	804	377
Tons of pig iron produced,.......	286,903	563,755
Rolling mills, bloomeries, and forges,	795	552
Tons of wrought iron produced,...	197,233	278,044

Out of the thirty-one States, at this period, ten had no blast furnaces, and twelve no works for the manufacture of wrought iron, although they contained large deposits of iron ore and some extensive coal fields. The day must, however, come when these will also take their places among the first in this branch of manufactures.

STATEMENT of Pig Iron produced in the United States in 1850, *together with the value of the production of the same article for* 1840, *the increase in ten years and the decrease in ten years.*

STATES.	Census of 1850.	Establishments.	Capital.	Raw Material Used.		Men Employed.	Average wages per month. Male.
				Tons of Ore.	Value.		
Alabama,	771,623	3	$ 11,000	1,838	$ 6,770	40	$17 60
Connecticut,	370,792	13	225,600	35,450	289,225	148	26 80
Delaware,	91,532	—	—	—	—	—	—
Georgia,	906,135	3	26,000	5,189	25,840	135	17 44
Illinois,	851,470	2	65,000	5,500	15,500	150	22 06
Indiana,	988,416	2	72,000	5,200	24,400	88	26 00
Kentucky,	982,405	21	924,700	72,010	260,152	1,845	20 23
Louisiana,	517,762	—	—	—	—	—	—
Maine,	583,169	1	214,000	2,907	14,939	71	22 00
Maryland,	583,034	18	1,420,000	99,866	560,725	1,370	20 14
Massachusetts,	994,514	6	469,000	27,909	181,741	263	27 52
Michigan,	397,654	1	15,000	2,700	14,000	25	35 00
Missouri,	682,044	5	619,000	37,000	97,367	334	24 28
New Hampshire,	317,976	1	2,000	500	4,900	10	18 00
New Jersey,	489,555	10	967,000	51,266	332,707	600	21 20
New York,	3,097,394	18	605,000	46,385	321,027	505	25 00
North Carolina,	869,039	2	25,000	900	27,900	26	8 00
Carried forward,	15,039,820	106	5,660,300	394,620	2,180,193	5,610	

STATEMENT of Pig Iron produced in the United States in 1850, *together with the value of the production of the same article for* 1840, *the increase in ten years, and the decrease in ten years—continued.*

STATES.	Census of 1850.	Establishments.	Capital.	Raw Material Used.		Men Employed.	Average wages per month. Male.
				Tons of Ore.	Value.		
Brought forward, ...	15,039,820	106	5,660,300	394,620	2,180,193	5,610	
Ohio, ...	1,980,329	35	1,503,000	140,610	630,037	2,415	24 48
Pennsylvania, ...	2,311,786	180	8,570,425	877,283	3,732,427	9,285	21 65
Rhode Island, ...	147,545	—	—	—	—	—	—
South Carolina, ...	668,507	—	—	—	—	—	—
Tennessee, ...	1,002,717	23	1,021,400	88,810	254,900	1,713	12 81
Vermont, ...	314,120	3	62,500	7,676	40,175	100	22 08
Virginia, ...	1,421,661	29	513,800	67,319	158,307	1,115	12 76
Wisconsin, ...	305,391	1	15,000	3,000	8,250	60	30 00
Total, ...	23,191,876	377	17,346,425	1,579,318	7,005,289	20,298	

STATEMENT of Pig Iron produced in the United States in 1850, *together with the value of the production of the same article for* 1840, *the increase in ten years, and the decrease in ten years—continued.*

STATES.	Annual Product.		1850.	1840.	Increase in Ten	Decrease in Ten
	Tons of Pig Iron.	Other Products.	Total Value.	Products.	Years.	Years.
Alabama, ...	522	$ 5,000	$ 22,500	$750	$21,750	—
Connecticut, ...	13,420	20,000	415,600	162,375	253,225	—
Delaware, ...	—	—	—	425	—	$425
Georgia, ...	900	28,000	57,300	12,350	44,950	—
Illinois, ...	2,700	—	70,200	3,950	66,250	—
Indiana, ...	1,850	—	58,000	20,250	37,750	—
Kentucky, ...	24,245	10,000	604,037	730,150	—	126,113
Louisiana, ...	—	—	—	35,000	—	35,000
Maine, ...	1,484	—	36,616	153,050	—	116,434
Maryland, ...	43,641	96,900	1,056,400	221,900	834,500	—
Massachusetts, ...	12,287	—	295,123	233,300	61,823	—
Michigan, ...	660	6,000	21,000	15,025	5,975	—
Missouri, ...	19,250	—	314,600	4,500	310,100	—
New Hampshire, ...	200	—	6,000	33,000	—	27,000
New Jersey, ...	24,031	—	560,544	277,850	282,694	—
New York, ...	23,022	12,800	597,920	727,200	—	129,280
North Carolina, ...	400	—	12,500	24,200	—	11,700
Carried forward, ...	168,612	177,800	4,128,340	2,655,275	1,919,017	455,952

STATEMENT of Pig Iron produced in the United States in 1850, *together with the value of the production of the same article for* 1840, *the increase in ten years, and the decrease in ten years—continued.*

STATES.	Annual Product.		1850.	1840.	Increase in Ten Years.	Decrease in Ten Years.
	Tons of Pig Iron.	Other Products.	Total Value.	Products.		
Brought forward,....	168,612	177,800	4,128,340	2,655,275	1,919,017	455,952
Ohio,................	52,658	—	1,255,850	880,900	374,950	—
Pennsylvania, ...	285,702	40,000	6,071,513	2,459,875	3,611,638	—
Rhode Island,.........	—	—	—	103,150	—	103,150
South Carolina, ...	—	—	—	31,250	—	31,250
Tennessee,............	30,420	41,900	676,100	403,213	272,887	—
Vermont, ...	3,200	—	68,000	168,575	—	100,575
Virginia,.............	22,163	—	521,924	470,262	51,662	—
Wisconsin, ...	1,000	—	27,000	75	26,925	—
Total, ...	563,755	259,700	12,748,727	7,172,575	6,257,079	680,927

STATEMENT of Iron Castings produced in the United States in 1850, together with the value of the production of the same article for 1840, the increase in ten years, and the decrease in ten years.

STATES.	Establishments.	Capital.	Tons of Pig Iron.	Value of Raw Material, Fuel, etc.	Men Employed.
Alabama, ...	10	$216,625	2,348	$102,085	212
Arkansas, ...	—	—	—	—	—
California, ...	1	5,000	75	8,530	3
Columbia, Dist. of, ..	2	14,000	545	18,100	27
Connecticut, ...	60	580,800	11,396	351,369	942
Delaware, ...	13	373,500	4,440	153,852	250
Georgia, ...	4	35,000	440	11,950	29
Illinois, ...	29	260,400	4,818	172,330	332
Indiana, ...	14	82,900	1,968	66,918	143
Iowa, ...	3	5,500	81	2,524	17
Kentucky, ...	20	502,200	9,731	295,533	558
Louisiana, ...	8	255,000	1,660	75,300	347
Maine, ...	25	150,100	3,591	112,570	243
Maryland, ...	16	359,100	7,220	259,190	761
Massachusetts, ...	68	1,499,050	31,134	1,057,904	1,596
Michigan, ...	63	195,450	2,494	91,865	337
Mississippi, ...	8	100,000	1,197	50,370	112
Missouri, ...	6	187,000	5,100	133,114	297
New Hampshire, ...	26	232,700	5,673	177,060	374
New Jersey, ...	45	593,250	10,666	301,048	803
New York, ...	323	4,622,482	108,945	2,393,768	5,925
North Carolina, ...	5	11,500	192	8,341	15
Ohio, ...	183	2,063,650	37,555	1,199,700	2,758
Pennsylvania, ...	320	3,422,924	69,501	2,372,467	4,782
Rhode Island, ...	20	428,800	8,918	258,267	800
South Carolina, ...	6	185,700	169	29,128	153
Tennessee, ...	16	139,500	1,682	90,035	261
Texas, ...	2	16,000	250	8,400	35
Vermont, ...	26	290,720	5,279	160,603	381
Virginia, ...	54	471,160	7,114	297,014	810
Wisconsin, ...	15	116,350	1,371	86,930	228
Total, ...	1,391	17,416,361	345,553	10,346,265	23,541

STATEMENT of Iron Castings produced in the United States in 1850, together with the value of the production of the same article for 1840, the increase in ten years, and the decrease in ten years—continued.

STATES.	Wages per month. Male.	1850. Products.	1840. Products.	Increase in Ten Years.	Decrease in Ten Years.
Alabama, ...	$30 05	$271,126	$27,700	$243,426	—
Arkansas, ...	—	—	1,240	—	$1,240
California, ...	23 33	20,740	—	20,740	—
Columbia, Dis. of,	27 05	41,696	68,000	—	26,304
Connecticut, ...	27 02	981,400	1,733,044	—	751,644
Delaware, ...	23 36	267,462	10,700	256,762	—
Georgia, ...	27 43	46,200	5,350	40,850	—
Illinois, ...	28 50	441,185	41,200	399,985	—
Indiana, ...	25 74	149,430	14,580	134,850	—
Iowa, ...	32 35	8,500	4,000	4,500	—
Kentucky, ...	24 89	744,316	164,080	580,236	—
Louisiana, ...	35 60	312,500	—	312,500	—
Maine, ...	29 00	265,000	56,512	208,488	—
Maryland, ...	27 50	685,000	312,900	372,100	—
Massachusetts, ..	30 90	2,235,635	1,798,758	436,877	—
Michigan, ...	28 68	279,697	57,900	221,797	—
Mississippi, ...	37 91	117,400	36,900	80,500	—
Missouri, ...	19 63	336,495	60,300	276,195	—
New Hampshire,	33 05	371,710	136,334	235,376	—
New Jersey, ...	24 00	686,430	405,955	280,475	—
New York, ...	27 49	5,921,980	2,512,792	3,409,188	—
North Carolina, .	23 46	12,867	16,050	—	3,183
Ohio, ...	27 32	3,069,350	784,401	2,284,949	—
Pennsylvania, ..	27 55	5,354,881	1,262,670	4,092,211	—
Rhode Island, ..	29 63	728,705	147,550	581,155	—
South Carolina, .	13 59	87,683	—	87,683	—
Tennessee, ...	17 96	264,325	100,870	163,455	—
Texas, ...	43 43	55,000	—	55,000	—
Vermont, ...	28 27	460,831	24,900	435,931	—
Virginia, ...	19 91	674,416	128,256	546,160	—
Wisconsin, ...	26 73	216,195	3,500	212,695	—
Total, ...		25,108,155	9,916,442	15,974,084	782,371

STATEMENT of Wrought Iron Manufactures produced in the United States in 1850, *together with the value of the production of the same article for* 1840, *the increase in ten years, and the decrease in ten years.*

STATES.	Establishments.	Capital.	Value of Raw Material.	Hands Employed—Male.	Average Wages per month. Male.	1850. Product.	1840. Product.	Increase in Ten Years.	Decrease in Ten Years.
Alabama, ...	3	$7,000	$3,355	34	$15 29	$7,500	$4,875	$2,625	—
Connecticut, ...	20	601,000	517,554	394	31 59	847,196	235,495	611,701	—
Delaware, ...	3	75,000	35,410	47	25 53	38,200	29,185	9,015	—
Georgia, ...	3	9,200	4,136	26	11 35	12,384	—	12,384	—
Indiana, ...	4	17,000	4,425	22	27 45	11,760	1,300	10,460	—
Kentucky, ...	4	176,000	180,800	183	32 06	299,700	236,405	63,295	—
Louisiana, ...	—	—	—	—	—	—	88,790	—	$88,790
Maryland, ...	17	412,050	386,216	468	24 31	771,431	513,500	257,931	—
Massachusetts, ..	58	2,561,100	2,430,533	2,472	29 46	3,908,952	390,260	3,518,692	—
Missouri, ...	2	42,100	24,509	101	30 00	68,700	7,670	61,030	—
New Hampshire,	3	7,000	11,575	9	31 34	20,400	8,125	12,275	—
New Jersey, ...	64	1,300,393	566,865	932	27 31	1,079,576	466,115	613,461	—
New York, ...	81	1,871,650	2,305,441	2,130	28 91	3,758,547	3,490,045	268,502	—
North Carolina, .	30	170,609	50,089	262	10 43	331,914	62,595	269,319	—
Ohio, ...	6	164,800	193,148	276	29 58	127,849	485,290	—	357,441
Pennsylvania, ..	162	7,828,916	5,698,563	6,591	28 31	9,224,256	5,670,860	3,553,396	—
Rhode Island, ...	2	209,400	112,123	222	27 85	223,650	—	223,650	—
South Carolina, .	—	—	—	—	—	—	75,725	—	75,725
Tennessee, ...	42	755,050	385,616	731	15 20	670,618	628,745	41,873	—
Vermont, ...	10	77,200	83,094	79	32 08	127,886	42,575	85,311	—
Virginia, ...	38	747,811	531,325	1,131	25 41	1,098,252	382,590	715,662	—
Total, ...	552	17,033,279	13,524,777	16,110		22,628,771	12,820,145	10,330,582	521,956

In 1851, the United States produced but 413,000 tons and imported 464,559 tons of all descriptions of iron, showing a falling off of more than one-half of the production of 1846–7. In 1852, the production of pig iron reached 540,755 tons, and the importation 501,158 tons, which was nearly one-half of the entire exports of Great Britain for that year.

Statement showing from what countries the deficiency of iron was supplied for the year 1852.

Imported from	Kind.	Tons.	Value.
Great Britain, ...	Pig, ...	91,149	$927,055
"	Old and Scrap, .	6,049	81,554
"	Castings, ...	351	18,114
"	Bar, ...	318,236	8,967,669
"	Steel, ...	8,550	1,629,222
"	Rods, hoop, sheet	19,125	789,140
"	Manufactures,	76,823	6,065,918
France, ...	Manufactures, ..	—	240,790
Sweden and Norway, .	Bar, ...	14,104	751,050
" "	Steel, ...	365	22,624
Russia, ...	Bar, ...	142	7,984
"	Sheet, ...	2,315	312,106
Belgium, ...	Manufactures, .	—	424,029
"	Others, ...	—	698
Other countries, ...	Manufactures, .	—	411,982
Total, ...	....	....	$20,661,592

This great increase in the imports of iron and steel must not, however, be attributed so much to the rapid increase of population, as to the combined effect of the free trade tariff of 1846 and the great influx of gold from California, which gave a stimulus to commerce and manufactures.

In the same year, Mr. Renton, of New Jersey, took out a patent for reducing the ores directly into malleable iron, which was afterwards carried out extensively in several of the States.

Comparative Statement of the Quarterly Price of Refined Bar Iron at the Ports of Boston, New York, Philadelphia, and Baltimore; with the Quarterly and Annual Average Price at the above four Ports for the last seventeen years.

PORTS.	1840.				1841.			
	January.	April.	July.	October.	January.	April.	July.	October.
Boston, ...	$101 00	$95 25	$89 75	$84 00	$81 25	$81 25	$78 50	$78 50
New York, ...	—	—	—	—	—	—	—	—
Philadelphia, ..	—	—	—	—	—	—	—	—
Baltimore, ...	—	—	—	—	—	—	—	—
Av'ge of 4 ports,	101 00	95 25	89 75	84 00	81 25	81 25	78 50	78 50
Yearly average,	—	—	—	92 50	—	—	—	79 87

PORTS.	1842.				1843.			
	January.	April.	July.	October.	January.	April.	July.	October.
Boston, ...	$71 25	$75 75	$67 25	$70 75	$69 25	$69 50	$64 50	$69 50
New York, ...	—	—	—	—	—	—	—	—
Philadelphia, ..	—	—	—	—	—	—	—	—
Baltimore, ...	—	—	—	—	—	—	—	—
Av'ge of 4 ports,	71 25	75 75	67 25	70 75	69 25	69 50	64 50	69 50
Yearly average,	—	—	—	71 25	—	—	—	68 19

PORTS.	1844.				1845.			
	January.	April.	July.	October.	January.	April.	July.	October.
Boston, ...	$67 50	$72 50	$70 75	$71 75	$78 25	$93 75	$81 50	$92 50
New York, ...	—	—	—	—	—	77 50	80 00	82 50
Philadelphia, ..	—	—	—	—	—	—	—	—
Baltimore, ...	—	—	—	—	—	—	—	—
Av'ge of 4 ports,	67 50	72 50	70 75	71 75	78 25	85 62	80 75	87 50
Yearly average,	—	—	—	70 62	—	—	—	83 03

Comparative Statement of the Quarterly Price of Refined Bar Iron at the Ports of Boston, New York, Philadelphia, and Baltimore; with the Quarterly and Annual Average Price at the above four Ports for the last seventeen years—continued.

PORTS.	1846.				1847.			
	January.	April.	July.	October.	January.	April.	July.	October.
Boston, ...	$88 00	$89 50	$86 00	$89 75	$82 75	$79 00	$79 75	$82 00
New York, ...	82 50	89 50	86 00	82 50	75 00	77 50	75 00	75 00
Philadelphia, ...	—	—	73 97	75 79	77 00	76 40	76 40	77 00
Baltimore, ...	—	—	80 00	75 00	70 00	75 00	72 50	72 50
Av'ge of 4 ports,	85 25	89 50	81 49	80 76	76 19	76 98	75 90	76 62
Yearly average,	—	— —	—	84 22	—	—	—	76 45

PORTS.	1848.				1849.			
	January.	April.	July.	October.	January.	April.	July.	October.
Boston, ...	$75 50	$68 50	$60 25	$60 25	$59 25	$62 25	$49 00	$48 00
New York, ...	75 00	68 00	59 00	60 00	60 00	60 00	48 00	49 00
Philadelphia, ...	63 75	64 36	56 24	56 84	55 43	58 45	47 81	47 17
Baltimore, ...	75 00	75 00	72 50	70 00	70 00	62 50	65 00	60 00
Av'ge of 4 ports,	72 31	68 96	62 00	61 77	61 17	60 80	52 45	51 04
Yearly average,	—	—	—	66 26	—	—	—	56 36

PORTS.	1850.				1851.			
	January.	April.	July.	October.	January.	April.	July.	October.
Boston, ...	$60 25	$55 75	$54 50	$51 25	$51 50	$51 00	$48 50	$47 75
New York, ...	47 50	45 00	43 50	51 00	47 00	40 00	40 00	42 00
Philadelphia, ...	48 78	48 38	46 81	46 57	44 96	47 17	46 57	46 57
Baltimore, ...	60 00	57 50	57 50	57 00	57 50	55 00	55 00	55 00
Av'ge of 4 ports,	54 13	51 66	50 38	51 58	50 24	48 29	47 52	47 83
Yearly average,	—	—	—	51 94	—	—	—	48 47

Comparative Statement of the Quarterly Price of Refined Bar Iron at the Ports of Boston, New York, Philadelphia, and Baltimore; with the Quarterly and Annual Average Price at the above four Ports for the last seventeen years—continued.

PORTS.	1852.				1853.			
	January.	April.	July.	October.	January.	April.	July.	October.
Boston,.... ...	$46 50	$46 00	$49 00	$60 50	$74 75	$72 00	$64 50	$74 50
New York,.....	43 00	41 50	47 75	60 50	81 00	80 00	63 00	71 00
Philadelphia,...	45 56	44 18	47 97	62 32	81 23	81 83	67 72	72 56
Baltimore, ...	52 50	52 50	52 50	60 00	80 00	87 50	85 00	80 00
Av'ge of 4 ports,	46 89	46 04	49 31	60 83	79 25	80 33	70 05	74 51
Yearly average,	—	—	—	50 77	—	—	—	76 03

PORTS.	1854.				1855.			
	January.	April.	July.	October.	January.	April.	July.	October.
Boston,........	$82 50	$80 50	$85 75	$85 50	$78 50	$64 25	$64 50	$74 50
New York,.....	80 00	82 50	82 50	81 00	65 00	58 00	62 50	67 50
Philadelphia,...	76 68	77 72	82 03	80 62	72 16	60 06	60 87	67 12
Baltimore, ...	77 50	82 50	85 00	87 50	87 50	82 50	75 00	75 00
Av'ge of 4 ports,	79 17	80 80	83 82	83 65	75 79	66 20	65 72	71 03
Yearly average,	—	—	—	81 86	—	—	—	69 68

PORTS.	1856.			
	January.	April.	July.	October.
Boston, ...	$72 50	$72 00	$70 50	—
New York, ...	66 25	70 00	65 00	—
Philadelphia, ...	—	—	—	—
Baltimore, ...	72 50	70 00	70 00	—
Average of four ports, ...	70 42	70 67	68 50	—
Yearly average, ...	—	—	69 86	—

STATEMENT of Prices of Steel (Duty paid) in New York, from 1851 *to* 1856, *inclusive.*

DESCRIPTION.	Whence Imported.	1851.	1852.	1853.	1854.	1855.	1856.
		Cents.	*Cents.*	*Cents.*	*Cents.*	*Cents.*	*Cents.*
Best quality cast steel,.........	England,	14 *a* 14½	14 *a* 14½	14 *a* 14½	15 *a* 15½	15½ *a* 16	14½ *a* 15
Second quality cast steel, ...	do.	12½ *a* 13	13 *a* 13½	13 *a* 13½	13 *a* 13½	13½ *a* 14	13 *a* 13½
Third quality cast steel,........	do.	11½ *a* 12	11 *a* 11½	11 *a* 11½	11 *a* 12	11 *a* 12	11½ *a* 12
Fourth qual. cast steel, machinery,	do.	9½ *a* 10	10 *a* —	10 *a* 10½	10½ *a* 11	10½ *a* 11	10 *a* 10½
Best quality shear steel,........	do.	14 *a* 14½	14 *a* 14½	14 *a* 14½	15 *a* 15½	15½ *a* 16	14½ *a* 15
Second quality shear steel, ...	do.	12½ *a* 13	13 *a* 13½	13 *a* 13½	13 *a* 13½	13½ *a* 14	13 *a* 13½
Best quality German steel, ...	Germany,	12 *a* 13	12½ *a* 13	12½ *a* 13	12½ *a* 13	12½ *a* 13	12½ *a* 13
do. do.	England,	10 *a* 10½	10 *a* 10½	10 *a* 10½	10½ *a* 11	10½ *a* 11	10½ *a* 11
Second quality German steel,...	do.	8 *a* 8½	8 *a* 8½	8 *a* 8½	8½ *a* 9	8½ *a* 9	8½ *a* 9
Third quality German steel, ...	do.	7 *a* 7½	7 *a* 7½	7 *a* 7½	7½ *a* 8	7½ *a* 8	7½ *a* 8
do. do.	Sweden,	5½ *a* 7½	5½ *a* 7½	5½ *a* 8	5½ *a* 8	5½ *a* 8	6½ *a* 8½
Fourth qual. Ger'n (spring) steel,	England,	4¾ *a* 5¼	4½ *a* 5	4¾ *a* 5¼	5½ *a* 6	4¾ *a* 5¼	5 *a* 5½
Best quality blister steel,.......	do.	12½ *a* 13	12½ *a* 13	12½ *a* —	12½ *a* 13	13 *a* 13½	12½ *a* 13
Second quality blister steel, ...	do.	10½ *a* 11	10½ *a* 11	10½ *a* 11	9½ *a* 10	10 *a* 10½	10½ *a* 11
Third quality blister steel,......	do.	8 *a* 8½	8 *a* 8½	8 *a* 8½	8½ *a* 9	9 *a* 9½	8½ *a* 9
Fourth quality blister steel, ...	do.	7 *a* 7½	7 *a* 7½	7 *a* 7½	7½ *a* 8	8 *a* 8½	7½ *a* 8
Milan steel, ...	Austria,	5¾ *a* 6	5¼ *a* 6	5 *a* 5½	5¼ *a* 5½	5¾ *a* —	7 *a* —
[do.	Sweden,	6	no sales.	5½ *a* —	5 *a* —	5½ *a* 6	no sales.

In 1853–4, the production of the United States was estimated at 805,000 tons. The actual cost of pig iron, with anthracite on the Lehigh, at this period, was from $15 to $17 per ton; while the cost of making it with charcoal, in several of the States, was from $25 to $28 per ton.

Statement of the Imports of Scotch Pig Iron into the United States in 1853 *and* 1854.

PORTS.	1853. Tons.	1854. Tons.
New York,............	84,378	60,186
Boston,...............	38,342	32,638
Philadelphia,..........	15,075	14,993
Baltimore,............	2,631	2,382
Providence,...........	3,755	6,242
New Orleans,.........	9,807	4,844
San Francisco,.........	50	40
Sundry ports,..........	1,800	2,548
Total,..............	155,857	113,873

In 1854, the long-predicted financial crisis arrived, which shook commercial credit to its foundation. The railroad enterprises, which had been pushed forward with so much vigor in this country, experienced a sudden check, and brought ruin upon many who had made large investments in them. The manufacturing industry of the country was also paralyzed, and the iron trade in particular suffered in common with all others. In Great Britain the crisis was also severely felt, and the manufacturers of malleable iron, who, a short time before, would scarcely have looked at *cinder pigs*, now used them largely, in the manufacture of iron, which was largely exported to the United

States and sold at a rate which crushed the American manufacturer.

In consequence of the rapid progress that railroads made in Europe and the United States, the price of rails advanced in England from £4 10*s*., in 1850, to £8 5*s*., in 1854, and the imports during that period exceeded all former years. Up to 1853, more than thirteen thousand miles of railways had been completed, and thirteen thousand more were projected and in the course of construction. In the haste, however, to build them, much inferior iron was imported, which proved a fruitful source of accidents and loss of life, as well as expenditure to relay them, which deprived stockholders of their dividends and bondholders of their interest. The depreciation of good rails should not exceed from five to eight per cent. per annum. On some of the best roads in the States, where careful examinations have been made and recorded, it has been found that the wear and tear of American rails have rarely exceeded that per centage, and it is impossible for the present capacity of American mills to supply the orders for them.

In the States west of the Alleghany mountains, where coal, limestone, and ores abound, and transportation by canal and rail is rapid and cheap, the American rail is fast superseding the English, not only on account of its costing less to deliver it at distant points in the West, but also of the superior quality of iron of which it is generally made. The following statement gives the cost of manufacturing pig iron in Ohio, in 1854, and the conversion of it into railroad bars.

Estimate of what it Cost to Manufacture Pig Iron in Ohio, in 1854.

Interest on $40,000 investment in furnaces, lands, buildings, etc.,	$4,000
Taxes, incidentals, repairs, etc.,	3,000
10,000 tons iron ore, at $2, to make 4,000 tons iron,	20,000
400,000 bushels coal, at 5 cts., " "	20,000
2,000 tons of limestone at $1 per ton,	2,000
Labor about the furnaces, at $2 per ton,	8,000
Handling, weighing, and transporting to rolling mill,	4,000
Cost of 4,000 tons pig at rolling mill,	$62,000

Or $15 50 per ton.

Estimate of Converting Pig Iron into Railroad Bars.

Interest on capital, $80,000 at ten per cent., 13,500 railroad bars, per ton,	$ 59
Wear and tear of machinery, and repairs, per ton,	1 50
Boiling furnace expenses per ton,	6 00
Rolling into muck rolls per ton,	50
Catching, hooking, dragging out, and shearing, per ton,	38
Piling, strapping, heating, rolling, catching, drawing, and straightening per ton,	3 00
Sawing, trimming, and straightening per ton,	60
Coal, 25 to 60 bushels, say 60 at 5 cts.,	3 00
Engine oil, miscellaneous, per ton,	1 00
Incidentals, weighing, draggage, and extras of the boiling furnace,	5 43
Cost of converting pig into rails, about	$22 00

This cost is based upon the supposition that the furnaces are separated from each other and from the rail mill; but that four blast furnaces and one rail mill belong to the same company. If the works were together, the cost would be about $37 per ton, allowing the ore to cost $2 per ton, and coal five cents per bushel. In many localities in Ohio, at the present price of labor, the cost would still be less, say $13 50 per ton for pig iron, and $34 to $36 per ton for railway bars. In Missouri, now one of the smallest of the iron-producing States, it has been estimated that there is ore enough, of the very best quality, within a few miles of the Pilot Knob * and Iron mountain, above the surface of the valleys in that State, to furnish fifty millions of tons per annum of manufactured iron, for the next two or three centuries; and, to work this inexhaustible quantity of ore, she can furnish one hundred millions tons of coal per annum, for the next five centuries. And, if you add to this the immense iron and coal resources of some of the other western States, no one can doubt that they will not only soon be able to manufacture all the iron required for home consumption, but also export large quantities to foreign countries.

* The Pilot Knob is a huge iron cone, rising from a plain and surrounded by mountains upon every side, its base being almost a circle, and its top terminating in a point or apex, like a sugar loaf. Its height is about five hundred and fifty feet, and its summit about fourteen hundred and seventy-two feet above tide water. The iron mountain covers a space of about five hundred acres, and its summit is about two hundred feet above the base. By a careful calculation, this mountain is estimated to contain 220,000,000 tons of iron ore above the base. The purity of the ore is as undoubted as its quantity, and yields about sixty-five per cent. of pure metal.

TABLE of Rail Mills in the United States, with their Capacities to Make, in 1854–5–6–7.

NAMES.	WHERE LOCATED.	1854.	1855.	1856.	1857.
Bay State,	Boston, Mass.,	15,000	15,000	16,000	17,871
Rensselaer,	Troy, N. Y.,	4,000	10,000	13,000	13,512
Trenton,	Trenton, N. J.,	10,000	10,000	13,000	16,000
Phœnix,	Phœnixville, Penn.,	13,688	14,500	15,000	18,592
Montour,	Danville, Penn.,	16,000	18,000	21,000	22,500
Rough and Ready,	Danville, Penn.,	4,500	5,000	5,000	5,500
Pottsville,	Pottsville, Penn.,	1,676	1,700	1,500	3,021
Lackawanna,	Scranton, Penn.,	10,982	18,000	16,000	11,338
Safe Harbor,	On Susquehanna, Penn.,	10,175	10,607	10,500	17,538
Mount Savage,	Cumberland, Md.,	7,000	7,500	12,000	7,357
Cambria,	Johnstown, Penn.,	1,806	11,000	8,000	7,159
Brady's Bend,	Brady's Bend, Penn.,	8,700	1,600	2,000	13,206
Washington,	Wheeling, Va.,	4,500	5,000	8,000	2,355
Covington,	Covington, Ky.,	—	—	—	1,976
Railroad Mill,	Cleveland, O.,	—	—	1,500	1,976
Newburg Mill,	Cleveland, O.,	—	—	—	1,800
Wyandotte,	Detroit, Mich.,	—	—	—	6,000
Gate City,	Atlanta, Ga.,	—	—	—	18,000
Palo Alto,	Pottsville, Penn.,	—	—	—	1,800
Newburg,	Newburg, N. Y.,	—	—	1,000	1,200

Other mills have also been recently built at St. Louis, Chicago, and Fairmount, near Philadelphia, Penn., which will probably increase the production thirty thousand tons, or more, of rails per annum.

Statement of the Iron Manufactures of Pennsylvania for the years 1854, 1855, *and* 1856.*

	1854.	1855.	1856.
	Tons.	Tons.	Tons.
Hot blast char. pig iron,..	73,554	66,970	108,100
Cold blast " " ..	54,329	56,225	79,300
Coke pig iron,..........	20,135	24,550	63,600
Raw bit's coal pig iron,..	14,684	12,500	18,900
Anthracite, " "	261,532	278,941	356,600
Total pig iron, ...	424,234	439,186	626,500
Blooms,...............	28,079	28,600	—
Hammered bars, ...	2,575	2,675	—
Sheets and plates, ...	20,408	21,505	—
Nails, rods, bars, etc., ...	104,535	121,550	—
Rails, ...	74,445	82,107	65,100

* "There were, in 1856, thirty-three forges in Philadelphia, Berks, Montgomery, and Chester counties. In Lancaster and York counties, fourteen; in eastern Pennsylvania, sixty-nine. The whole number was one hundred and sixteen, forty-seven of which belong to the trade of Philadelphia. Six within the city use steam-power, as do two at Reading, one at Weissport, and one at Tamaqua—all others using water-power. Those within the city produced, in 1856, nearly 1,700 tons of blooms; with about 1,050 tons of axles, shafting, and finished work. Within the radius of fifty miles, the amount was about 5,600 tons of blooms, and 2,100 tons of axles and shafting—exclusive of the works of the great Reading steam forge, at which the shafts of the Adriatic were made a year or two since. In eastern Pennsylvania the product was about 28,000 tons of blooms, and 5,500 tons of shafting, bars, etc.

"There are nine large rolling mills belonging directly to the city, inclusive of the Cheltenham rolling mill just north of the city limits, and of the Pencoyd mill just across the Schuylkill, at the mouth of the Wissahicon.

"There are twenty-seven rolling mills within fifty miles of Philadelphia; three at Conshohocken, thirteen miles distant on the Schuylkill; three at Norristown; four at Phœnixville; seven near Coatesville, in Chester county; five at or near

In 1855, the consumption of iron in the United States amounted to 1,310,000 tons, showing an increase of more than two hundred per cent. in ten years, a ratio altogether beyond the increase of any other great interest in the country.

From numerous trials, under every conceivable variety of circumstances, it is now conceded that iron made from the Lake Superior ores, is at least equal if not superior to the best Swedish or Russian.

The following table shows the comparative quality,

Reading, etc. Their consumption in 1856 was as follows, estimating 20,000 tons for the Phœnixville mills, and one or two others not obtained.

American pig iron,	34,000	tons.
Blooms,	11,010	"
Scrap iron,	2,050	"

The production was as follows, for 1856:

Railroad iron (rails),	18,600	tons.
Boiler and other plate,	12,300	"
Bars, nails, axels,	14,800	"

The Phœnixville mills alone, of this division, roll railroad iron.

"In other parts of eastern Pennsylvania, there are twenty-seven additional rolling mills; three at or near Pottsville; one at Weissport; one at Scranton; three at Danville; four at and near Harrisburg; three in Lancaster county; two near Williamsport; five near Bellefonte, etc. Two at Pottsville, the Scranton mill, the Danville mills, and the Safe Harbor mill, seven in all, make rails; producing, in the aggregate, 46,500 tons of rails in 1856.

"In other items they make—

Boiler and other plate,	2,190	tons.
Bars, rods, and nails,	15,500	"

"The iron consumed cannot be given for all these establishments. The Montour rail mills used 27,300 tons of pig iron in making 22,800 tons of rails; the Lancaster rail mill, 9,800 tons of pig iron to make 7,350 tons of rails. The rail mills are immense establishments; that at Phœnixville making 18,600 tons of rails; the Montour mills (three) making 22,800 tons; and the Scranton mill 11,400 tons, all in 1856.

"The whole amount of railroad iron made in eastern Pennsylvania was thus 65,100 tons for 1856. The whole amount imported during 1856–7 (the fiscal year) was, according to the Treasury report, 179,305 tons."

and gives the results of the various experiments of Professor Johnson on the tensity of American, English, Swedish, and Russian bar iron :

		Number of Trials.	Strength per lb. Square Inch.
Iron from	Salisbury, Conn.,	40	58,600
"	Centre county, Penn.,	4	59,400
"	McIntyre Essex county, N.Y.,	4	59,962
"	Lancaster county, Penn.,	5	58,661
"	Sweden,	4	58,184
"	England (cable bolt), E.V.,	5	59,105
"	Russia,	5	76,069
"	Carp River, Michigan, U.S.,	1	89,582

Already the raw ores and pig metal of Lake Superior begin to form an important item in the trade of that lake. In 1848, two years after the discovery of these mines, the first bloom forge was built, which is now worked by the Cleveland Iron Mining Company. This was followed by another, built by the Marquette Iron Company, in 1850. The third bloomery was built in 1853 ; and the fourth by the Collins Iron Company in 1855, three miles from Marquette, which is capable of manufacturing two thousand tons of blooms per annum. But little of the ore, however, could be brought to market until the completion of the Sault Ste. Marie Ship Canal, when the following shipments were made to various ports on the Lakes :

1855,	1,447 tons.
1856,	11,597 "
1857,	26,375 "

The great superiority of this ore for manufacturing purposes has also created an extensive demand for it on the Atlantic seaboard, as well as in the western

States, where extensive iron works have been erected to make bar, pig, and rolled iron. The iron produced from it is soft and fibrous, extremely malleable and remarkable for its quality, and capability of being converted into cast steel.

STATE ASSAYER'S OFFICE, BOSTON, *Sept.* 13, 1856.

I have made a chemical analysis of the samples of iron on section given T 27 R 27, near Marquette, Lake Superior, and find it to consist of the finest and best of iron, and like that of the Dannamora iron ore of Sweden, capable of making the finest kind of iron. It yields, by analysis, the following ingredients, *per cent.* :

Peroxide of iron,	98.02
Oxide of manganese,	1.28
Silica,	0.44
Lime,	0.32
	100.06

The gain of 6.100 of a grain is from absorption of oxygen by the small proportion of protoxide in the ore, which, in analysis, is converted into the peroxide ; oxide of manganese, in the proportion found in this ore, is very beneficial, and will render the iron capable of making the *best* of steel. I find no titanium, phosphorus, sulphur, arsenic, chrome, or other injurious substances in this ore, *and am able to certify that it is the very best kind of iron ore known in the world,* and will *make good cast iron in the furnace,* or fine bloom and bar in the forge. The ore contains 68,644 per cent. of pure iron, but the practical yield in the furnace will be somewhat less, as some oxide of iron is lost in the cinders and slag as the ore comes from the mine ; it will probably yield about 60 per cent. of cast iron in the blast furnace.

(Signed) C. T. JACKSON, M.D.
Assayer, etc.

Extract from the State Assayer's Office, Paris.

"Ecole des Mines," No. 2001.

Specimens of Iron Ore from the neighborhood of Marquette, Lake Superior, Mich., given by G. W. Barr :

No. 1. St. Clair and Smith Mountain. Oxide of iron, somewhat micaceous. Not magnetic.

No. 2. St. Clair and Smith Mountain. Anhydrous peroxide of iron. Compact with magnetic ore.

No. 3. St. Clair and Smith mountain. Oxide of iron. Slightly micaceous. Not Magnetic.

No. 4. Near Marquette. Oxide of iron in grain and sheety, magnetic ore.

No. 5. Eureka mine. Red oxide of iron with some ochre.

No. 6. Cleveland Mountain. Specular ore—crystaline.

No. 7. Slaty or steel ore from Jackson and Cleveland Mountains. Specular ore with quartz.

No. 8. Jackson Mountain. Specular ore ; crystaline. Same appearance as No. 6.

No. 9. Fine granular ore of Cleveland and Jackson Mountains. Specular ; compact.

No. 10. Jackson Mountain. Red oxide of iron.

No. 11. One-half jasper, one-half iron ore. Jackson Mountain. Specular ore ; somewhat micaceous ; compact.

No. 12. Found in a quarry at dock. Carbonate of lime and iron.

No. 13. Red oxide of iron with specular ore.

None of the above specimens contain *phosphorus*, *arsenic* or *sulphur*. The *gangue* is a mixture of *quartz* and of *alumina*, of *oxide of iron*, of *lime*, and of *alkalies*. No. 7 contains some silicate of iron.

(Signed), L. E. Rivot,

Professor of Analytical Chemistry and Director of the Assay Office, Paris.

Table of the Analysis of Lake Superior Iron Ores.

Number.	Gangue.	Metal Iron.	Lime.	Oxygen and Loss.	Alumina.	Magnesia.	Alkalies.	Oxide of Iron.	Oxide of Magnesia.	Carbonic Acid.	Water.	Soluble Silica.	Total.
1	17.00	58.00	0.20	24.80	—	—	—	—	—	—	—	—	100.00
2	28.50	49.00	1.00	21.50	—	—	—	—	—	—	—	—	100.00
3	41.00	40 00	0.80	18.20	—	—	—	—	—	—	—	—	100.00
4	12.50	58.00	2.00	27.50	—	—	—	—	—	—	—	—	100.00
5	14.50	58.00	2.50	25.00	—	—	—	—	—	—	—	—	100.00
6	4.00	65 00	1.70	39.30	—	—	—	—	—	—	—	—	100.00
7	6 00	67 00	1.00	26.00	—	—	—	—	—	—	—	—	100.00
8	8.50	64.00	0.20	27.30	—	—	—	—	—	—	—	—	100.00
9	28.00	50.00	0.50	21.50	—	—	—	—	—	—	—	—	100 00
10	44.50	32.50	2.50	20.50	—	—	—	—	—	—	—	—	100.00
11	44.50	38.50	1.00	16.00	—	—	—	—	—	—	—	—	100.00
12	6.00	—	30.00	—	0.50	5.30	0.80	7.00	1.30	31.50	11.20	6.40	100.00
13	25.50	51.00	0.50	23.00	—	—	—	—	—	—	—	—	100.00
14*	4.50	67.00	Tra's	28.50	—	—	—	—	—	—	—	—	100.00

* Nearly the same as No. 6.

Many attempts have been made to manufacture steel directly from the ore, but without much success. In most of the cities of the United States, steel is made by the English method. In eastern and western Pennsylvania, upwards of six thousand tons were manufactured in 1849.

Several patents have been taken out both in England and this country which have considerably reduced the cost of its manufacture and increased its production. Among these, the Riepe and the Uchatius in Great Britain, and the Neville patent in this country, have become the most celebrated.

The Riepe patent consists in a peculiar method of working in the puddling furnace; in converting pig iron or alloys of pig iron and wrought iron into steel, with the coöperation of clay in the furnace, and by the coöperation of atmospheric air. "He employs," says Mr. Clay, of the Mersey Steel Works, "the puddling furnace in the same way as for making wrought iron, and introduces a charge of about 280 lbs. of pig iron, and raises the temperature to redness. As soon as the metal begins to fuse and trickle down in a fluid state, the damper is to be partially closed, in order to temper the heat. From 12 to 16 shovelsful of iron cinder discharged from the rolls or squeezing machine are added, and the whole is to be uniformly melted down. The mass is then to be puddled, with the addition of a little black oxide of manganese, common salt, and dry clay, previously ground together. After this mixture has acted for some minutes, the damper is to be fully opened, when about 40 lbs. of pig iron are to be put into the furnace, near the fire bridge, upon elevated beds of cinder prepared for that purpose. When this pig iron begins to trickle down, and the mass on the bottom of the furnace begins to boil and throw out from the surface the well-known blue jets of flame, the said pig iron is raked into the boiling mass, and the whole is then well mixed together. The mass soon begins to swell up, and the small grains begin to form in it and break through the melted cinder on the surface. As soon as these grains appear, the damper is to be three-quarters shut, and the process closely inspected while the mass is being puddled to and fro beneath the covering layer of cinder. During the whole of this process, the heat should not be raised above cherry redness, or the welding heat of shear steel. The blue jets of flame gradually disappear, while the formation of grains continues, which grains very soon begin to fuse together, so that the mass becomes waxy, and has the above-mentioned cherry redness. If these precautions be not observed, the mass would pass more or less into iron, and no uniform steel product could be obtained. As soon as the mass is finished so far, the fire is stirred to keep the necessary heat for the succeeding operation ; the damper is to be entirely shut, and part of the mass is collected into a ball, the remainder always being kept covered with cinder slack. This ball is brought under the hammer, and then worked into bars. The same process is continued until the whole is worked into bars. When pig iron made from sparry iron ore, or mixtures of it with other pig iron, is used, he only adds about 20 lbs. of the former pig iron at the latter period of the process, instead of about 40 lbs. When he employs Welsh or pig iron of that description, he throws 10 lbs. of best plastic clay, in a dry granulated state, before the beginning of the process, on the bottom of the furnace, and adds, at the latter period of the process, about 40 lbs. of pig iron, as before, but strews over it clay in the same proportion. He does not claim the commencement of this process for making steel in the puddling furnace, but the regulating the heat in the finishing process, and excluding the atmospheric air from the mass in the manner as described, and also the use or addition of iron to the mass towards the latter part of the process.

"The balls, instead of being rolled into bars, may be hammered into slabs or blooms, for such uses as forgings, rails, plates, or any hammered or rolled steel which requires to be perfectly solid; but for ordinary use, puddled bars are made, at the Mersey Iron Works, from 2 to 14 inches wide, which are afterwards cut up and piled for various purposes. In using the puddled bar steel, it has been found very desirable to test each bar before using it, and to closely inspect the quality, and to select such as is best adapted to the purposes required; for instance, for steel rails, or railway points, or switches, which Mr. Clay rolls at one operation direct to the regular taper form desired, under a patent which he has 'for rolling iron or other metal of taper form,' he selects the most crystalline steel for the upper and under surfaces of the rail or switch, and for the interior that which is of a more fibrous and tougher description. Between the centre and top and bottom of the rail he places steel of an intermediate grade, which causes the whole pile or mass to weld up easily and work solid. It is necessary in this, as in any operation in which steel is used, to take the greatest possible care in the heating and working of the material, but from the first commencement there has been found no difficulty in heating, rolling, or forging this steel into any form or shape, as it has been made into steel plates, bars, angle steel, rivet steel, rails, railway points, and forgings of all kinds, with perfect ease and success; and ever since the manufacture was commenced at the Mersey Steel and Iron Works, this steel has been used for almost everything that was required to be of a strong and durable nature, or to repair any of those breakages which are of such constant occurrence in every iron-work.

"It is somewhat worthy of remark that, although this process is so novel, and, apparently, of so delicate a nature, yet, with the specification as his only guide, having never before heard or seen the operation, it succeeded perfectly in the first trial which was made, and produced so excellent a steel, that after working about 100 tons, it has hardly been surpassed. He had used pig iron of all descriptions, North Welsh, South Welsh, Staffordshire, and Scotch, with the same result—the production of an excellent steel; but he had not found, so far, anything like the great difference that he expected between hot and cold-blast iron. Most excellent results have been obtained from both; this is more particularly important as it shows that the extent to which this manufacture may be carried need not be circumscribed by the very limited supply of cold-blast pig iron.

"The puddled-steel bar when broken shows a clear crystalline and even fracture, and has the usual sonorous and musical tone when struck. The crystals appear much finer and more regular than in the ordinary blister steel; in fact, to the unpractised eye the appearance was quite like that of the best cast steel, and it has all those distinguishing features by which steel is known from iron. It hardens to any degree that may be requisite, taking all the colors which develop themselves under the different degrees of heat, and may be made into such articles as ordinary chisels direct from the puddling bar; it will take a very fine polish, and has the same amount of elasticity that steel usually possesses.

"One extraordinary feature in regard to this wrought steel is, that it can be produced either of a harsh, hard, unyielding character, or of a soft, silky, fibrous structure, or of any of the grades between these two points, and that a bar when quite cold may be bent up double and perfectly close (with extreme difficulty certainly, on account of the great stiffness of the material) without the slightest sign of fracture, but, when forced back again, a beautiful long silky fibre is apparent; or if a piece of steel plate be partly cut through with a chisel and then broken, it appears beautifully fibrous; if made into a tool, for instance, and hardened, it at once assumes the crystalline character peculiar to steel. Referring to cast steel as a material for ordnance, Mr. Clay imagines that the want of elasticity complained of may be partially accounted for thus: Cast steel requires a very high temperature to render it fluid for founding, which necessarily causes a considerable amount of shrinking in the casting when passing from the fluid to the solid state, and the casting is of that peculiar crystalline structure which is produced under such conditions (weakened to a great extent also by the strain caused by shrinkage), unless the steel casting is afterwards subjected to the hammering or rolling process before mentioned, by which the particles of steel are relieved from their shrinking strain, and are consolidated and allowed to assume a comparative state of repose. In the manufacture of forgings from puddled steel, the case is very different. We possess, in the best puddled steel, as great, if not a greater amount of strength, as in cast steel under the most favorable circumstances; and as the particles of wrought or puddled steel are never in a state of fusion from the time of their first formation in the puddling furnace, the enormous contractile strain incident upon the transition of the steel from the fluid to the solid state is avoided in the first place, and also the grain of the puddled steel may be so placed in the forging to be made as the strain which it will be called upon to resist may require, and the different descriptions of steel, whether crystalline or fibrous, may be arranged in the best positions as regards strength and durability.

"Steel forgings have been made at the Mersey Steel and Iron Works into piston rods (some with the piston forged solid, 18 in. diameter, for a nasmyth hammer), large roll screws, shear pins of all sorts, rolls for rolling iron, hammers and anvils, and for a variety of other purposes. In making these forges no difficulty was experienced; rather more time was required on account of the necessity of heating the steel slowly, and also because the hammer did not make the same impression on it that it does upon iron. The effect of forging upon this steel is to consolidate it, and when broken in the usual manner, the appearance of the crystals is much finer than when it is rolled, as might be expected. Of all the various uses to which this steel may be applied, there is perhaps none so important as its application to marine and railway purposes; for the former use the material offers directly so considerable a saving in regard to weight, with an equal amount of strength (putting out of the question its durability and other advantages), that its universal adoption can hardly be doubted."

By the Uchatius process, pig iron is granulated by pouring it while in a molten state into water, which is then in the best state for conversion into cast steel. To the pig iron thus reduced, pulverized sparry iron and fine clay are added; or grey oxide of manganese used instead of the latter. This mixture is then put into crucibles, and the process of melting and casting is proceeded with in the usual way for producing cast steel. In order to obtain harder kinds of steel, charcoal is added in small quantities to the above-mentioned combinations.

The patent obtained in the United States, by John Neville, of New York, is entirely successful. The basis of this invention consists in the introduction into crucibles, along with the pieces of wrought or malleable iron, of certain chemicals in which cyanogen is contained. The usual furnaces and melting pots suitable for melting blister steel may be employed. The malleable iron is prepared by breaking it up into small pieces. In a twenty pounds charge of iron in a crucible are introduced ten ounces of charcoal; six ounces of common table-salt; half an ounce of brick dust, or oxide of manganese; one ounce of sal ammoniac; and a half ounce of ferrocyanide of potassium. The pot is then covered and introduced into the furnace, and the contents thoroughly melted, the heat being maintained for three or four hours. The mass is then poured off into moulds, in the ordinary way of pouring off cast steel, and with the usual care required for producing a solid ingot. This may be then rolled into sheets, or hammered and tilted into bars, after the common method.

If a particular quality of steel and temper is required, it will only require a proper understanding of the chemicals to produce it, and each batch can be duplicated to any extent, producing an exact uniformity, which is one of the most essential points gained in making it by this process.

The steel manufactured by this process costs less than by any other known method, although, probably, the minimum cost price has not yet been reached. The great abundance of magnetic ore in this country, and the advantageous situation of some of its beds near the sea coast, offer great facilities for the manufacture of steel, which will render, in a very few years, the United States independent of foreign countries.

Next to Michigan, there is no State in the Union where magnetic ores, capable of yielding the finest iron and steel, are more profusely distributed than in New York; and it is a matter of the greatest national importance, that they should be more extensively worked. Already they have, to some extent, been shipped to Pennsylvania, New Jersey, and Maryland, and manufactured into steel, which has been pronounced equal to any imported. The following tables show the composition of most of the ores of that State, and the proportion of metallic iron they contain.

TABLE exhibiting the Composition of several Iron Ores found in the State of New York, and the proportion of Metallic Iron which they contain, in 100 parts.

	1.	2.	3.	4.	5.	6.	7.	8.
Protoxide of Iron, …… Peroxide of Iron, ……	25.40 70.50	96.50	44.10 52.75	95.75	98.90	89.00	94.84	42.26 53.69
Oxide of Manganese, ….	1.60	—	traces.	—	—	—	—	—
Silica, …………… Alumina, ……………	and loss. 2.50	3.50	3.15	4.25	1.10	11.00	and titanic acid. 5.16	4.05
Titanic Acid, ……….	—	—	traces.	—	—	—	—	—
Metallic Iron, ………..	69.10	71.11	71.22	70.55	72.86	65.57	69.98	70.13

No. 1. Magnetic oxide of iron. Long Mine, Orange county.
" 2. " " Rich Iron Mine, Monroe, Orange county.
" 3. " " Forsher Mine, " "
" 4. " " Oneil Mine, " "
" 5. " " Wilks Mine, " "
" 6. " " Denny Mine, Philipstown, Putnam county.
" 7. Iron Sand Banks of Moon River, Lewis county.
" 8. Magnetic oxide of iron. Fort Ann, Washington county.

TABLE exhibiting the Composition of several Iron Ores found in the State of New York, and the proportion of Metallic Iron which they contain, in 100 parts—continued.

	9.	10.	11.	12.	13.	14.	15.	16.
Protoxide of Iron, ... }	70.80	92.97	24.50	92.15	27.00	98.00	97.00	96.52
Peroxide of Iron, ... }			66.80		71.50			
Oxide of Manganese, ...	—	—	—	—	—	—	traces.	—
Silica, ... }	and loss. 1.20	5.93	8.70	and other insoluble matter. 7.85	and loss. 1.50	2.33	2.83	3.48
Alumina, ... }								
Titanic Acid, ...	—	—	—	—	—	—	—	—
Carbonate of Lime, ...	25.00	—	—	—	—	—	—	—
Plumbago, ...	3.00	—	—	—	—	—	—	—
Lime, ...	—	—	—	—	—	traces.	traces.	—
Metallic iron, ...	52.16	68.50	65.81	67.89	71.00	68.60	67.90	67.56

No. 9. Magnetic oxide of iron. Warrensburg, Warren county.
" 10. " " Penfield Mine, Crownpoint, Essex county.
" 11. " " Crag Harbor, Essex county.
" 12. " " Newcomb, "
" 13. " " Arnold Mine, Clinton county.
" 14. Light blue ore. " "
" 15. Grey ore. " "
" 16. Specular oxide of iron. Kearny Mine, St. Lawrence county.

TABLE exhibiting the Composition of several Iron Ores found in the State of New York, and the proportion of Metallic Iron which they contain, in 100 parts—continued.

	17.	18.	19.	20.	21.	22.	23.	24.
Peroxide of Iron, ...	96.00	51.50	42.93	72.37	80.27	82.90	71.00	68.00
Oxide of Manganese, ...	—	—	—	—	—	traces.	—	8.50
Silica, ... }	4.00	6.00	17.99	15.43	7.43	3.60	8.50	6.50
Alumina, ... }		7.15						
Carbonate of Lime, ...	—	24.50	28.33	—	—	—	—	—
Carbonate of Magnesia, ..	—	7.75	10.40	—	—	—	—	—
Water, ...	—	and loss. 2.75	—	12.20	11.66	13.50	20 50	17.00
Metallic Iron, ...	67.20	36.05	30.05	50.65	56.18	58.03	49.70	47.60

No. 17. Specular oxide of Iron. Pierpont, St. Lawrence county.
" 18. Lenticular clay iron ore. Wolcott, Wayne county.
" 19. " " Near Rochester, Monroe county.
" 20. Limonite. Southfield, Richmond county.
" 21. Hematite. Union Vale, Dutchess county.
" 22. " Amenia, "
" 23. Bog iron ore. St. Lawrence county.
" 24. Brown iron ore. Monroe, Orange county.

It is not only in the production, but in the application of iron to new purposes, or in the extension of its use in others, that marks the progress of the United States in the last half century. Railway iron, locomotive and marine engines, iron bridges, ships, and buildings, are now among the most prominent industrial pursuits of this country; and in no branch of American manufactures has more skill been shown, or greater progress made, than in the manufacture of hardware, cutlery, and edge tools.

The progress of iron manufactures in the western States, is truly wonderful, and continues unabated. The consumption of pig iron in these States was estimated, in 1857, at over three hundred thousand tons, of which Pittsburg consumed more than one-half in her manufactures. In that city there are twenty-five iron and steel rolling mills which consume—

	Value.
105,333 tons pig iron,	$3,159,990 00
27,267 " blooms,	2,181,360 00
4,931 " scrap iron,	186,440 00
2,550 " Swedes and rolled iron,	178,500 00
6,187,515 bushels of coal,	251,500 60
118,000 " coke,	5,900 00
5,040 tons fire clay,	21,500 00
2,095,000 fire brick,	41,900 00
9,258 tons ore,	120,696 00
51,800 gallons oil and grease,	53,034 00
Small items to amount of	43,000 00
Total,	$6,243,820 60
They employ 4,623 hands, whose yearly wages amount to	$2,366,020 00
The capital in the ground, buildings, and machinery, employed in prosecution of the business, is	$3,280,000 00

They produce as follows:

		Value.
3,212½	tons boiler iron,	$ 388,712 00
67,100	" bar, of various sizes,	4,697,000 00
5,637	" sheet iron,	681,077 00
699,762	kegs nails, spikes, and rivets,	2,797,048 00
10,000	boxes tacks,	50,000 00
800	tons galvanized and imitation Russia iron,	96,000 00
10,850	" blister, plow, spring, and cast steel,	1,747,850 00
2,500	crow bars,	5,000 00
1,500	sledges,	1,875 00
Axles to amount of		80,500 00
Springs "		135,000 00
Vices "		50,000 00

Sixteen foundries which consume—

		Value.
19,200	tons of pig iron,	$576,000
540,500	bushels coke and coal,	27,025
81,000	fire brick,	1,620
79,500	bushels fire clay,	11,720
1,266	barrels blacking,	3,798
1,575	gallons oil,	1,732
Lumber to amount of		7,000
Iron and nails, "		14,500
Hardware, "		2,325
Sand, "		2,320
Loam, "		2,700
Total,		$650,740

They employ 860 hands, whose yearly wages are, ...$346,500

They keep twenty steam engines running, and the capital in the grounds, buildings, and machinery, is.. 498,000

They produce 16,890 tons of castings of the various descriptions, before enumerated, worth.........1,248,300

In these foundries may be daily seen cast, articles ranging from the heavy Columbiad (cannon) weighing

15,300 pounds, throwing a ball of 124 pounds, to the finest articles of Berlin work.

Besides the above establishments, there are other factories and machine-shops which consume about twenty-five thousand tons of pig iron, per annum. From 1842 to 1855, upwards of 1,600 cannon were cast, and 25,000 tons of shell and shot for cannon and howitzers. In the manufacture of steam engines, upwards of 4,000 tons of wrought, and 12,000 tons of pig iron are now annually consumed. The total value of the iron manufactures of Cincinnati, in 1857, was estimated at nine millions of dollars.

Table of the Exports and Imports of Cincinnati, from 1846 *to* 1856.

Year.	Exports of Manufactured Iron.	Imports of Iron and Steel.	Imports of Pig Iron.	Imports of Nails.
	Tons.	Tons.	Tons.	Kegs.
1846, ...	1,238	29,165	13,685	56,130
1847, ...	5,646	34,213	15,868	59,983
1848, ...	6,916	29,889	21,145	55,893
1849, ...	6,270	55,158	15,612	57,546
1850, ...	5,767	66,817	17,211	83,173
1851, ...	9,776	63,903	16,110	83,761
1852, ...	11,329	54,178	23,604	64,189
1853, ...	14,246	78,568	30,179	76,541
1854, ...	18,322	96,713	41,807	85,391
1855, ...	17,978	113,456	26,613	106,350
1856, ...	11,881	136,870	41,016	174,560

The rolling mills, last year, produced iron manufactures valued at $3,645,000, and the foundries $3,712,000.

The following tables exhibit the population, importation, exportation, production and home consumption of iron manufactures, with the allotment per capita thereof, of the United States, for the years 1840, 1850, and an estimate for 1855.

STATEMENT exhibiting the value of the foreign importations and exportations, domestic exportations, home consumption of foreign importations, and home consumption of foreign importations less the domestic exportations of iron and manufactures of iron and iron and steel; also the foreign importations and exportations, home consumption of foreign importations, total home consumption of foreign iron and manufactures of iron and iron and steel, and foreign cast, shear, German, and other steel, the total home consumption of foreign iron and manufactures of iron and iron and steel, and foreign cast, shear, German, and other steel, less the domestic exportations; also the manufacture of pig iron, iron castings, wrought iron, and the manufactures thereof in the United States, total manufacture of pig iron, iron castings, and wrought iron, and the manufactures thereof in the United States, consumption of domestic iron, and the manufactures thereof, total consumption of foreign and domestic iron, and the total consumption of foreign and domestic iron and manufactures of iron; also cast, shear, German, and other steel, in the United States, for the years 1840 *and* 1850, *with an estimate thereof for* 1855, *on the same ratio of increase as between the years* 1840 *and* 1850.

YEARS.	IRON, AND MANUFACTURES OF IRON AND IRON AND STEEL.					CAST, SHEAR, GERMAN, AND OTHER STEEL.			Total home consumption of foreign iron, and manufactures of iron and iron and steel, and foreign cast, shear, German and other steel.
	Foreign importations.	Foreign exported.	Domestic exportations.	Home consumption of foreign importations.	Home consumption of foreign importations less domestic exportations.	Foreign importations.	Foreign exported.	Home consumption of foreign importations.	
1840,.......	$6,750,099	$156,115	$1,104,455	$6,593,984	$5,489,529	$528,716	$33,961	$494,755	$7,088,739
1850,.......	16,333,145	100,746	1,911,320	16,232,399	14,321,079	1,332,253	40,193	1,292,060	17,524,459
1855,.......	22,980,728	1,565,523	3,753,472	21,415,205	17,661,733	2,593,137	63,068	2,530,069	23,945,274

STATEMENT exhibiting the value of the foreign importations and exportations, domestic exportations, home consumption of foreign importations, and home consumption of foreign importations less the domestic exportations of iron and manufactures of iron and iron and steel; also the foreign importations and exportations, home consumption of foreign importations, total home consumption of foreign iron and manufactures of iron and iron and steel, and foreign cast, shear, German, and other steel, the total home consumption of foreign iron and manufactures of iron and iron and steel, and foreign cast, shear, German, and other steel, less the domestic exportations; also the manufacture of pig iron, iron castings, wrought iron, and the manufactures thereof in the United States, total manufacture of pig iron, iron castings, and wrought iron, and the manufactures thereof in the United States, consumption of domestic iron and the manufactures thereof, total consumption of foreign and domestic iron, and the total consumption of foreign and domestic iron and manufactures of iron; also cast, shear, German, and other steel, in the United States, for the years 1840 *and* 1850, *with an estimate thereof for* 1855, *on the same ratio of increase as between the years* 1840 *and* 1850*—continued.*

YEARS.	Total home consumption of foreign iron, and manufactures of iron and iron and steel, and foreign cast, shear, German, and other steel, less domestic exportations.	Manufacture of pig iron in the United States.	Manufacture of iron castings in the United States.	Manufacture of wrought iron, and the manufactures thereof, in the United States.	Total manufacture of pig iron, iron castings, and wrought iron, and manufactures of wrought iron, in the United States.	Consumption of domestic iron, and the manufactures thereof, in the United States.	Total consumption of foreign and domestic iron, and the manufactures thereof, in the United States.	Total consumption of foreign and domestic iron, and manufactures of iron; also cast, shear, German, and other steel in the United States.
1840,	$5,984,283	$7,172,575	$9,916,442	$12,820,145	$29,909,162	$28,804,707	$35,398,691	$35,893,446
1850,	15,613,139	12,748,727	25,108,155	22,628,771	60,485,653	58,574,333	74,806,732	76,098,792
1855,	20,191,802	16,016,910	34,012,021	28,377,607	78,406,538	74,653,066	96,068,271	98,598,340

STATEMENT exhibiting the population, production of pig iron, iron castings, and manufactures of wrought iron, with the allotment per capita thereof; the consumption of domestic iron and the manufactures thereof, with the allotment per capita; the home consumption of foreign importations of iron, and manufactures of iron and iron and steel, and cast, shear, German, and other steel, with the allotment per capita; and the total consumption of foreign and domestic iron, and manufactures of iron and iron and steel, cast, shear, German, and other steel, in the United States, and the allotment per capita thereof, for the years 1840, 1850, *and an estimate for* 1855.

YEARS.	Population.	Total product of pig iron, iron castings, wrought iron, and manufactures of wrought iron, in the United States.	Allotment per capita of the product of pig iron, iron castings, wrought iron, and manufactures of wrought iron, in the United States.	Consumption of domestic iron, and the manufactures thereof, in the United States.	Allotment per capita of the value of the consumption of domestic iron, and the manufactures thereof, in the United States.	Home consumption of foreign importations of iron, and manufactures of iron and steel, and cast, shear, German, and other steel.	Allotment per capita of home consumption of foreign importations of iron, and manufactures of iron and steel, and cast, shear, German, and other steel.	Total consumption of foreign and domestic iron, and manufactures of iron and steel; also cast, shear, German, and other steel, in the United States.	Allotment per capita of the consumption of foreign and domestic iron, and manufactures of iron and steel; also cast, shear, German, and other steel, in the United States.
1840, ...	17,069,453	$29,909,162	$1 75.22	$28,804,707	$1 68.75	$7,088,739	$0 43.53	$35,893,446	$2 10.28
1850, ...	23,191,876	60,485,653	2 60.80	58,574,333	2 52.56	17,524,459	75.56	76,098,792	3 28.12
1855, ...	27,185,517	78,406,538	2 88.45	74,653,066	2 74.60	23,945,274	88.08	98,598,340	3 62.68

STATEMENT exhibiting the yearly value of iron, and manufactures of iron, and iron and steel, cast, shear, German, and other steel, imported from and exported to foreign countries; domestic exports of like articles; home consumption of foreign iron, and manufactures of iron, and iron and steel; home consumption over the domestic exports of the same articles, and the total consumption of foreign iron, manufactures of iron, and iron and steel, cast, shear, German, and other steel, over domestic exportations, for the last seventeen years, and the yearly average for the aforesaid period.

YEARS.	IRON, AND MANUFACTURES OF IRON, AND IRON AND STEEL.				
	Foreign, imported.	Foreign, exported.	Domestic, exported.	Home consumption of foreign iron and manufactures of iron, and iron and steel.	Home consumption of foreign iron, and manufactures of iron, and iron and steel, over domestic exportations.
1840, ...	$6,750,099	$156,115	$1,104,455	$6,593,984	$5,489,529
1841, ...	8,914,425	134,316	1,045,264	8,780,109	7,734,845
1842, ...	6,988,965	177,381	1,109,522	6,811,584	5,702,062
1843, ...	1,903,858	50,802	532,693	1,853,056	1,320,363
1844, ...	5,227,484	107,956	716,332	5,119,528	4,403,196
1845, ...	8,294,878	91,966	845,017	8,202,912	7,357,895
1846, ...	7,835,832	122,587	1,151,782	7,713,245	6,561,463
1847, ...	8,781,252	63,596	1,167,484	8,717,656	7,550,172
1848, ...	12,526,854	98,295	1,259,632	12,428,559	11,168,927
1849, ...	13,831,823	109,439	1,096,172	13,722,384	12,626,212
1850, ...	16,333,145	100,746	1,911,320	16,232,399	14,321,079
1851, ...	17,306,700	100,290	2,255,698	17,206,410	14,950,712
1852, ...	18,957,993	134,937	2,303,819	18,823,056	16,519,237
1853, ...	27,255,425	262,343	2,499,652	26,993,082	24,473,430
1854, ...	29,341,775	795,872	4,210,350	28,545,903	24,335,553
1855, ...	22,980,728	1,565,523	3,753,472	21,415,205	17,661,733
1856, ...	22,041,939	423,221	4,161,008	21,618,718	17,457,710
Yearly average,	13,839,598	264,434	1,830,804	13,575,164	11,744,360

Nothing is better calculated to show the material progress of the iron trade of the United States, than the above and following statements, notwithstanding the blighting effect of the free-trade principles upon its production. In 1846, the production of pig iron was estimated at 765,000 tons, yet, in 1856, it had only reached 782,958 tons, showing an increase of only 17,958 tons, in face of a consumption of more than a million and a half of tons, the deficiency being supplied principally by Great Britain, without cheapening the price to the consumer.

STATEMENT exhibiting the yearly value of cast, shear, German, and other steel, imported from and exported to foreign countries; domestic exports of like articles; and the total consumption of foreign cast, shear, German, and other steel, over domestic exportations, for the last seventeen years, and the yearly average for the aforesaid period.

YEARS.	CAST, SHEAR, GERMAN, AND OTHER STEEL.			Total home consumption of foreign iron, and manufactures of iron, and iron and steel, and foreign cast, shear, German, and other steel.	Total home consumption of foreign iron, and manufactures of iron, and iron and steel, and foreign cast, shear, German, and other steel over domestic exportations, for the last seventeen years.
	Foreign, imported.	Foreign, exported.	Home consumption of foreign cast, shear, German and other steel.		
1840, ...	$528,716	$33,961	$494,755	$7,088,739	$5,984,283
1841, ...	609,201	24,848	584,353	9,364,462	8,319,198
1842, ...	597,317	18,447	578,870	7,390,454	6,280,932
1843, ...	201,772	59,733	142,039	1,995,095	1,462,402
1844, ...	487,462	15,415	472,047	5,591,575	4,875,243
1845, ...	775,675	20,052	755,623	8,958,535	8,113,518
1846, ...	1,234,408	32,564	1,201,844	8,915,089	7,763,307
1847, ...	1,126,458	19,218	1,107,240	9,824,896	8,657,412
1848, ...	1,284,937	41,397	1,243,540	13,672,099	12,242,467
1849, ...	1,227,138	55,044	1,172,094	14,894,478	13,798,306
1850, ...	1,332,253	40,193	1,292,060	17,524,459	15,613,139
1851, ...	1,570,063	38,371	1,531,692	18,738,102	16,482,404
1852, ...	1,703,599	31,569	1,672,030	20,495,086	18,191,267
1853, ...	2,970,313	31,637	2,938,676	29,931,758	27,432,106
1854, ...	2,477,709	53,247	2,424,462	30,970,365	26,760,015
1855, ...	2,593,137	63,068	2,530,069	23,945,274	20,191,802
1856, ...	2,538,323	25,598	2,512,725	24,131,443	19,970,435
Yearly average,	1,368,146	35,551	1,332,595	14,907,759	13,076,955

In conclusion, the iron production of the United States may be classed under three different heads:

1. The blast furnaces, using either anthracite, charcoal, raw or coked bituminous coal.

2. The bloomeries or mountain forges, which turn ore or cast iron into blooms or malleable iron.

3. Rolling mills, which convert these into bar, rod, sheet, and nail plate iron.

Statement showing the entire production of Pig Iron in the United States, in 1854, 1855, *and* 1856.

FURNACES.	1854.	1855.	1856.
Anthracite Furnaces.	Tons.	Tons.	Tons.
In Pennsylvania,	208,703	255,326	306,966
Out of Pennsylvania,	99,007	87,779	41,573
Charcoal and Coke Furnaces.			
East Pennsylvania,	62,724	60,596	51,775
N. W. Pennsylvania,	78,927	59,388	59,587
S. W. Pennsylvania,	11,052	18,217	29,400
Charcoal Furnaces.			
East of the Hudson,	30,420	30,926	27,837
Northern and western New York,	19,197	19,736	18,847
Southern N. York and N. Jersey,	13,435	7,901	5,683
Maryland,	35,658	36,309	30,998
N. Western Virginia,	1,930	2,342	1,467
Eastern and middle Virginia,	5,880	6,926	5,730
North and South Carolina,	1,820	1,830	1,926
Georgia and Alabama,	3,604	3,682	4,302
Tennessee,	38,596	30,000	30,000
Missouri,	5,213	6,000	13,201
West Kentucky,	5,000	5,000	5,000
East Kentucky,	22,830	15,580	21,160
S. Ohio (charcoal and coke),	56,081	47,182	69,605
N. Ohio ditto,	8,289	6,025	7,901
Illinois, Indiana, Michigan, Wisconsin, and Minnesota,	5,000	5,000	50,000
Total tons,	713,366	705,745	782,958

www.ingramcontent.com/pod-product-compliance
Lightning Source LLC
LaVergne TN
LVHW011219110826
845150LV00006B/1474

* 9 7 8 1 4 2 5 5 1 6 4 5 1 *